Presenting research from a wide range of scientists working on intertidal sediments – both mud flats and saltmarshes – this book is of importance to all environmental scientists.

The individual chapters explore the underlying biogeochemical processes controlling the behaviour of carbon, the nutrients nitrogen and phosphorus, and contaminants such as toxic organics, trace metals and artificial radionuclides in intertidal environments. The biogeochemistry of these environments is critical to understanding their ecology and management. All of the chapters include both a comprehensive review and the results of recent research. The authors are active researchers and the book brings together their different perspectives on this diverse and ecologically important environment.

This book is designed for researchers and managers working on intertidal environments, but it will also serve as a valuable senior undergraduate and graduate reference text in environmental chemistry, environmental science, earth science and oceanography.

BIOGEOCHEMISTRY OF INTERTIDAL SEDIMENTS

CAMBRIDGE ENVIRONMENTAL CHEMISTRY SERIES

Other books in the series:
P. Brimblecombe *Air Composition and Chemistry Second Edition*
A.C. Chamberlain *Radioactive Aerosols*
M. Cresser and A. Edwards *Acidification of Freshwaters*
M. Cresser, K. Killham and A. Edwards *Soil Chemistry and its Applications*
R.M. Harrison and S.J. de Mora *Introductory Chemistry for the Environmental Sciences Second Edition*
S.J. de Mora *Tributyltin: Case Study of an Environmental Contaminant*

Biogeochemistry of intertidal sediments

Edited by

T.D. JICKELLS AND J.E. RAE

CAMBRIDGE UNIVERSITY PRESS
Cambridge, New York, Melbourne, Madrid, Cape Town, Singapore, São Paulo

Cambridge University Press
The Edinburgh Building, Cambridge CB2 2RU, UK

Published in the United States of America by Cambridge University Press, New York

www.cambridge.org
Information on this title: www.cambridge.org/9780521483063

First published 1997
This digitally printed first paperback version 2005

A catalogue record for this publication is available from the British Library

Library of Congress Cataloguing in Publication data

Biogeochemistry of intertidal sediments / [edited by] T.D. Jickells and
J.E. Rae.
p. cm. – (Cambridge environmental chemistry series; 9)
Includes index.
ISBN 0 521 48306 9
1. Marine sediments. 2. Biogeochemistry. I. Jickells, T.D.
(Tim D.) II. Rae, J.E. (Joy E.) III. Series.
GC380.15.B55 1997
551.46′083–dc20 96-38998 CIP

ISBN-13 978-0-521-48306-3 hardback
ISBN-10 0-521-48306-9 hardback

ISBN-13 978-0-521-01742-8 paperback
ISBN-10 0-521-01742-4 paperback

Contents

Contributors

K. Carpenter, Hydraulics Research, Wallingford, Howbery Park, Wallingford, Oxfordshire, OX10 8BA, UK

M.L. Coleman, Postgraduate Research Institute for Sedimentology, University of Reading, PO Box 227, Whiteknights Park, Reading, RG6 2AB, UK

W.M. Duan, Postgraduate Research Institute for Sedimentology, University of Reading, PO Box 227, Whiteknights Park, Reading, RG6 2AB, UK

T.D. Jickells, School of Environmental Studies, University of East Anglia, Norwich, NR4 7TJ, UK

S.R. Jones, The Princess Royal Building, Westlakes Research Institute, Westlakes Science and Technology Park, Moor Row, Cumbria, CA24 3LN, UK

S.J. Malcolm, Ministry of Agriculture, Fisheries and Food, Directorate of Fisheries Research, Pakefield Road, Lowestoft, Suffolk, NR33 0HT, UK

P. McDonald, The Princess Royal Building, Westlakes Research Institute, Westlakes Science and Technology Park, Moor Row, Cumbria, CA24 3LN, UK

K. Pye, Postgraduate Research Institute for Sedimentology, University of Reading, PO Box 227, Whiteknights Park, Reading, RG6 2AB, UK

J.E. Rae, Postgraduate Research Institute for Sedimentology, University of Reading, PO Box 227, Whiteknights Park, Reading, RG6 2AB, UK

S.J. Rowland, Petroleum and Environmental Chemistry Group, Department of Environmental Sciences, University of Plymouth, Drake Circus, Plymouth, PL4 8AA, UK

G. Ruddy, School of Environmental Studies, University of East Anglia, Norwich, NR4 7TJ, UK. (Present address Plymouth Marine Laboratory, Citadel Hill, Plymouth, PL1 2PB, UK)

D.B. Sivyer, Ministry of Agriculture, Fisheries and Food, Directorate of Fisheries Research, Pakefield Road, Lowestoft, Suffolk, NR33 0HT, UK

A. Turner, Department of Environmental Sciences, University of Plymouth, Drake Circus, Plymouth, PL4 8AA, UK

A.O. Tyler, BMT Port & Coastal Limited, Grove, House, 5 Ocean Way, Ocean Village, Southampton, SO1 1TJ, UK

J.L. Zhou, Petroleum and Environmental Chemistry Group, Department of Environmental Sciences, University of Plymouth, Drake Circus, Plymouth, PL4 8AA, UK

Preface

This book arises from a meeting of the same name held at the Postgraduate Research Institute for Sedimentology, Reading, UK, in April 1994. The meeting was sponsored by the Challenger Society and the Geochemistry Group of the Mineralogical Society: we are grateful to both these organisations for their generous support. Eight scientists active in research on the biogeochemistry of intertidal systems were invited to present papers on their own particular interests within the field. We deliberately selected a mixture of established and young UK scientists and invited them to both briefly review the field and to present some new data from their own research. These papers have now been formalised into this book together with an overview chapter written by ourselves. The various authors have taken different approaches to the task we set them: some have achieved a balance between review and current research, whereas others have chosen to concentrate mainly on either review or recent research. As editors we have encouraged their diversity, allowing the authors the freedom to deal with the subject as they see best. We wish to thank all the authors for their good humour and hard work. All the chapters have been peer reviewed so we are additionally grateful to D. Cooper, A. Grant, J. Hamilton-Taylor, D. Hydes, S. Malcolm, R. Mortimer, J. Scudlark and S. Wakefield. (Chapter one was reviewed by the authors of the other book chapters.)

The work presented here demonstrates that despite the diversity of intertidal habitats there are common biogeochemical principles in operation. It is our hope that the book might enhance the appreciation and understanding of these complex and beautiful environments and that in so doing it might contribute to their preservation.

This book is University of Reading PRIS Contribution No 441.

1

○ ○ ○ ○ ○ ○ ○ ○ ○ ○ ○ ○ ○ ○ ○ ○ ○ ○ ○ ○

Biogeochemistry of intertidal sediments

T.D. Jickells and J.E. Rae

In some places the boundary between land and sea is in the form of abrupt and often spectacular cliffs but, elsewhere, the boundary can take the form of a complex environment of intertidal sediments. These environments include shingle banks, sandy beaches, mud flats, saltmarsh and mangrove (or mangal) communities. In some cases one or other of these environments will occur, in others they will be associated with one another. For example, on many North Sea and North American East Coast shorelines, mud flats grade into saltmarshes behind the shelter of shingle spits and sand dunes. In general, saltmarshes and mangroves occupy similar ecological niches with mangroves at lower latitudes (winter temperatures greater than 10 °C) and saltmarshes at high latitudes, though in some locations both communities coexist (Chapman, 1977).

The considerable global significance of these intertidal systems is evident from Figure 1.1. Around the coast of Britain alone there are 44 370 hectares (ha) of saltmarsh (Allen & Pye, 1992) and 589 429 ha on the US East Coast (Reimold, 1977) with a total global area of 3.8×10^7 ha (Steudler & Peterson, 1984 and references therein). There are 365 500 ha of mangal forest around the Indian subcontinent and another 250 000 ha in the Mekong delta (Blasco, 1977). The total global area of mangal is about 2.4×10^7 ha (Twilley, Chen & Hargis, 1992). The total area of intertidal sediments is likely to be similar to that of adjacent saltmarshes and mangals.

These intertidal environments afford very effective coastal protection (Brampton, 1992). The importance of such coastal protection is increasing as a result of two processes. Firstly, the growth in world population is increasing the pressure on land in general. Moreover, the increase in populations in coastal areas is disproportionately large, compounding this pressure (Hinrichsen, 1994). Secondly, the process of sea level rise is threatening low lying land with increased risk of flooding. Sea level rise is

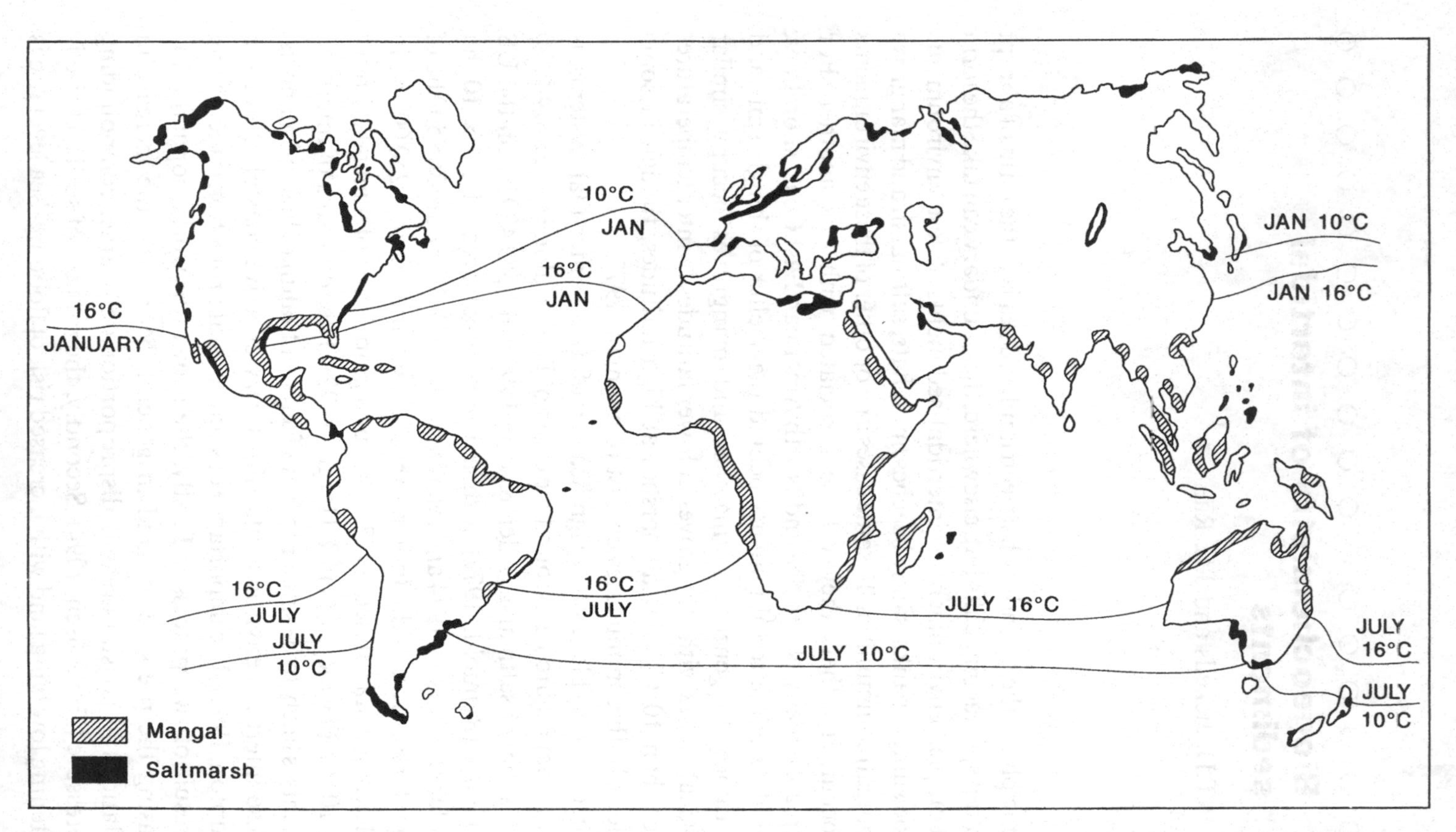

Figure 1.1 Map showing global distribution of saltmarshes and mangroves (mangal) in relation to average temperature, redrawn from Chapman, 1977.

in part a natural process, as the coastal system slowly responds to the end of the last glaciation, but is now being exacerbated by anthropogenically induced climate change – the greenhouse effect (e.g. Wigley & Raper, 1992). The threat to coastal areas from climate change arises not just from sea level rise associated with global warming, but also potentially from changes in weather patterns and hence wave and current patterns (Tooley, 1992).

In addition to providing a valuable sea defence, intertidal sediments offer an important habitat for wildlife, food and recreation. The very nature of intertidal areas has left them relatively undisturbed by human activity compared to inland areas. This coupled with the rich food supply in the muddy sediments means that many intertidal areas are now very important wildlife sanctuaries and nursery grounds for fish and invertebrates (Adam, 1990). Thus, for example, up to 12 million birds of 50 different species live for at least part of the year on the vast shallow water or intertidal muds of the Wadden Sea off the north Dutch and German coast (North Sea Task Force, 1993). Beside these quantifiable environmental roles as wildlife habitats and coastal defences, these intertidal areas have an intrinsic beauty that has always attracted people.

For all these reasons, intertidal sedimentary environments are important areas for scientists to study, and there is a long history of such work. This has often focused on the geomorphology of these areas and how this can be used to interpret the geological record (e.g. Nummedal, Pilkey & Howard, 1987; Allen and Pye, 1992) and the ecology of these areas (e.g. Adam, 1990; Chapman, 1977; Mathieson and Nienhuis, 1991). In addition many studies have concentrated on one type of ecosystem (e.g. saltmarsh or mangal). This book aims to take a different and complementary approach. Firstly, it focuses on the chemistry of these systems as modulated by and interacting with the geological and biological environment – hence the term biogeochemistry. Secondly, we have drawn no distinction between the different types of intertidal ecosystem since we believe the fundamental biogeochemical processes are the same in all the systems, though of course the final chemical system observed by our measurement varies as a function of the ecosystem. Thus the biogeochemical environment of a mangal swamp and an arctic mud flat may be very different, but the fundamental principles regulating these environments will be similar. Some of the authors in this book have even drifted out into more open waters to illustrate their points; something that only serves to emphasise both the generality of the biogeochemical principles and also the limited research effort expended into some aspects of the biogeochemistry of intertidal areas.

The detrital mineral phases of the intertidal sediments (e.g. clays and quartz) are of limited geochemical interest in the context of this book. Rather it is the organic matter and the chemicals adsorbed *to* the mineral phases of the sediment and to one another that drive most of the geochemical processes. Adsorption to sediments is a function of sediment surface area and hence particle size, with the finest sediments adsorbing the most material, as illustrated in the chapters by McDonald & Jones, Zhou & Rowland, and Rae. This means that in a geochemical (and also an ecological) sense the coarse sediments of high energy environments like shingle and sand beaches are of limited interest. The fine-grained organic matter which feeds the flora and fauna and fuels the geochemical reactions is washed out of these areas, to accumulate in the low energy environments such as the mud flats, mangals and saltmarshes that are the main focus of this book. Hence, the sunbathers can keep the sandy beaches and leave the geochemists to wallow in the mud!

These areas of fine sediment accumulate in both estuarine and open coast environments, though in the latter case they usually form in areas of reduced tidal and wave energy, such as behind barrier island complexes or in embayments (Allen & Pye, 1992). As noted earlier, in such environments there may be a continuum from sand dune through saltmarsh to mud flats, and the relative sizes of these can vary greatly. The marshes of Sapelo Island Georgia (USA) occupy 75% of the available coastal lagoon (Wiegert, Pomeroy & Wiebe, 1981) and in the 'big swamp' phase (6000 yr BP) many Australian estuaries were almost completely filled with mangal forest, to be replaced more recently by a mixture of mangal and mud flats (Woodroffe, 1990). By contrast in the sheltered embayment of the Wash coast in England (Malcolm & Sivyer, this volume) about 4000 ha of saltmarsh (Doodey, 1992) are fronted by about 30 000 ha of intertidal mud flats (S.J. Malcolm, pers. comm.).

These areas of sediment accumulation are transitory features on a whole range of time scales. They have migrated dramatically over the Holocene period (last 10 000 years) as relative sea levels have risen globally due to deglaciation, though this pattern of sea level change has varied locally (Allen & Pye, 1992). This is illustrated for the North Sea in Figure 1.2, and similar changes occurred in other coastal areas. As noted earlier, increases in sea level over the next 100 years may accelerate as a result of the greenhouse effect. Intertidal systems have responded to these changes by migrating where this is practical or accreting sediment sufficiently quickly to retain their position relative to sea level (e.g. Funnell & Pearson, 1989; Tooley, 1992). Maximum marsh vertical accretion rates appear to be 10–20 mm y^{-1}, similar to maximum predicted rates of sea level rise, so, provided sediment supply is maintained (which may depend on policies of coastal defence in sediment supply regions), marshes should be able to keep pace with sea level rise (Boorman, Goss-Custard & McGrorty, 1989; Reed, 1990). Because of the longer life span and slower growth rates of mangrove trees, mangals may not be able to respond so readily (Woodroffe, 1990).

Predicting the exact response of the whole intertidal sedimentary ecosystem to sea level rise is extremely difficult, though progress is being made, often with surprising results. Thus in the Wadden Sea, a rise in sea

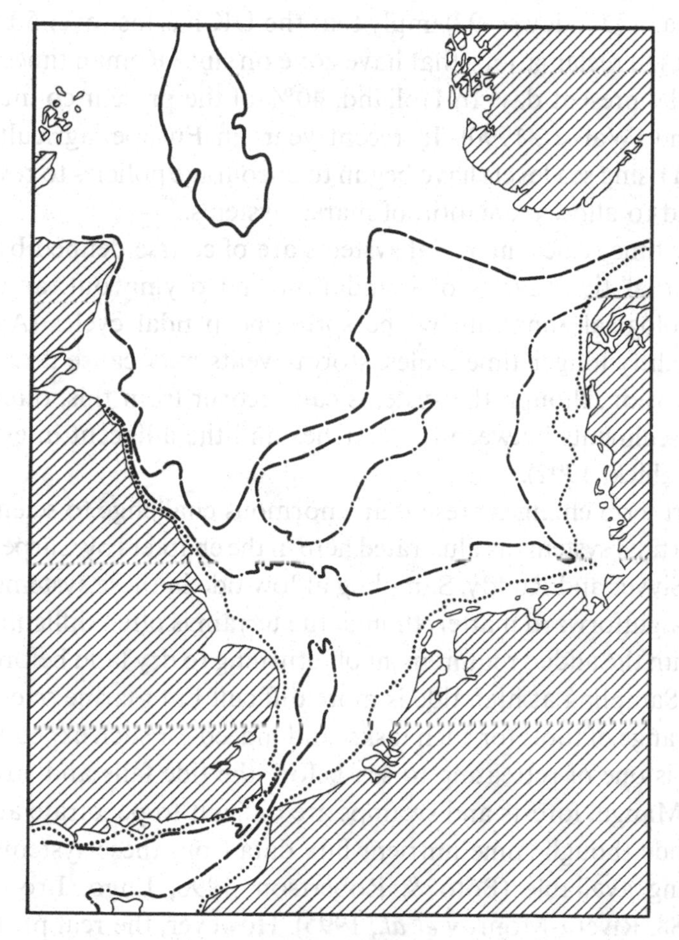

Figure 1.2 Map showing estimated locations of shore lines in the North Sea from 18 000 years ago to present, based on Jelgersma, 1979. ——— 18 000 yr BP, sea level 130 m below present day, — — — 10 300 yr BP, sea level 65 m below present day, —·—·—·— 8700 yr BP, sea level 36 m below present day, ... 7800 yr BP, sea level 20 m below present day.

level may result in increased exposure of intertidal areas and hence increased primary production (Peerbolte, Eysink & Ruardij, 1991). In areas where sediment supply is not maintained, loss of the intertidal ecosystems will occur. This is evident in areas such as the Nile and Mississippi deltas where damming has dramatically reduced sediment supply. This, combined with local subsidence, has meant that sea level rise exceeds sedimentation rates and contributes to marsh deterioration, at least in Louisiana (Reed, 1990), but probably also in the Nile delta, and indirectly in the Indus delta (Halim, 1991).

Other human impacts on intertidal systems are less subtle with a long history of drainage and protection from inundation to yield agricultural land. The data of Doodey (1992) imply that the UK has lost over 50% of its saltmarshes to reclamations that have gone on since Roman times and continue to the present day. In Holland, 40% of the present country is reclaimed land (Walker, 1990). In recent years in Europe, agricultural surpluses and rising sea levels have begun to encourage policies to reverse this trend and to allow recreation of marsh systems.

On shorter time scales intertidal systems are of course defined by the dramatic diurnal tidal cycles of inundation and drying out, and the saltmarsh ecological zonation by the spring/neap tidal cycles (Adam, 1990). On slightly longer time scales, storm events may cause dramatic short term erosion, though the systems can recover from this, often by exchange of sediments between the marshes and the adjacent intertidal mud flats (Pethick, 1992).

These short term changes present an enormous challenge to scientists studying intertidal systems as illustrated here in the chapters by Carpenter, Malcolm & Sivyer and Ruddy. Sampling at low tide in these systems is a relatively straightforward matter, though the terrain is often difficult and dangerous with the added requirement of returning to dry land before the tide returns. Sampling at high tide is more difficult but marine scientists have a wide range of tools for such tasks, and indeed for such expeditions the problem is one of retreating to sea before the tide falls and strands researchers. Mangal forests and swamps are difficult to penetrate at any state of the tide though some nutrient flux data from these systems are now becoming available (Boto & Robertson, 1990; Lugo, Brown & Bronson, 1988; Rivera-Monroy *et al.*, 1995). However, the real problem for scientists working in intertidal systems is coping with the dramatic and rapid changes from a system under water to one above water, and one where water flows reverse and change on very short time scales. The contrasting approaches of Carpenter, Ruddy and Malcolm & Sivyer in this volume illustrate two of the available strategies. These approaches are designed to address changes associated with diurnal tidal rhythms and also spring/neap tidal cycles. Pethick (1992), French & Spencer (1993) and Nyman, Crozier & de Laune (1995) present evidence that storm events may dominate sedimentation on saltmarshes. Such events are extremely difficult to study by direct observations because of their unpredictability, as well as the dangers of working in these environments under such conditions. We therefore have no direct measurements of fluxes during storm events, though they can be indirectly estimated (e.g. Nyman, Crozier & de Laune, 1995). Therefore we must be extremely cautious in extrapolating data from relatively low energy situations to such high energy conditions.

The primary productivity of many of these intertidal communities is high relative to offshore communities – 200–500 gC m^{-2} yr^{-1} for saltmarshes and mangroves (Mann, 1982) and about 100 gC m^{-2} yr^{-1} for benthic algae on mud flats (Cadée & Hegeman, 1974). This high productivity is supported in part by very high rates of N_2 fixation. Capone and Carpenter (1982), for example, estimate that almost half of all marine N_2 fixation takes place in saltmarshes and mangroves, though more recent data (e.g. Howarth *et al.*, 1988; Boto & Robertson, 1990) suggests substantially lower rates. Very little of the productivity of marshes and mangroves appears to be grazed while alive, but is rather decomposed as detritus by micro-organisms (Mann, 1982; Adam, 1990) partly in or on the sediments, together with the products of benthic algal production, and partly after export to adjacent coastal waters. Based on studies of carbon isotopes in particulate organic matter in saltmarsh creeks, this export appears to be minor (Adam, 1990) though the subject of considerable debate (Carpenter, this volume). For nitrogen, even if net exchanges are minor, transformations may occur such as from dissolved to particulate forms (e.g. Rivera-Monroy *et al.*, 1995) or nitrate to ammonium (Carpenter, this volume). Thus much of this plant detritus, together with decomposing below-ground plant material and organic carbon associated with imported fine sediment, provides the sediments with a rich supply of organic matter. This organic matter ultimately feeds the invertebrates and birds of these areas and also drives the geochemical reactions in the sediments.

It is the decomposition of this organic matter which provides one of the unifying biogeochemical concepts in all these intertidal environments. In all areas subject to fine-grained sedimentation, organic matter content of sediments is relatively high. This organic matter is subsequently oxidised by bacteria using a variety of oxidising agents or terminal electron acceptors (TEA) which are themselves reduced, hence the term redox reactions. These different TEAs yield different amounts of energy from their reactions with organic carbon (Table 1.1), so the microbial community using the highest energy yielding system always dominates, assuming there is a significant amount of that TEA present. Thus, as long as there is oxygen present, this will always be the oxidising agent used, but, once this is exhausted, alternative oxidants are used in a fixed sequence beginning with nitrate (Coleman, 1985; Ruddy, this volume). It is for this reason that these intertidal sediments can consume large amounts of nitrate very quickly (Malcolm & Sivyer, this volume). In many systems nitrate levels are naturally low, but increasing concentrations of nitrate in rivers arising from human activity can mean that nitrate is an important electron acceptor in some coastal systems today. Once nitrate is exhausted in the sediments the readily available oxidised iron and manganese are reduced to their more soluble reduced form. Since phosphate in sediments is usually strongly associated with iron, this iron mobilisation can also

Table 1.1 Redox equilibria (based on Andrews *et al.*, 1995)

Aerobic respiration
$O_2 + CH_2O \rightarrow CO_2 + H_2O$
$\Delta G = -29.9$ kcal/mol

Denitrification
$\frac{4}{5}NO_3^- + CH_2O = \frac{4}{5}H^+ \rightarrow \frac{2}{5}N_2 + CO_2 + \frac{7}{5}H_2O$
$\Delta G = -28.4$ kcal/mol

Manganese reduction
$2MnO_2 + CH_2O + 2HCO_3^- + 2H^+ \rightarrow 2MnCO_3 + 3H_2O + CO_2$
$\Delta G = -23.3$ kcal/mol

Nitrate reduction
$\frac{1}{2}NO_3^- + CH_2O + H^+ \rightarrow \frac{1}{2}NH_4^+ + CO_2 + \frac{1}{2}H_2O$
$\Delta G = -19.6$ kcal/mol

Iron reduction
$4FeOOH + 4HCO_3^- + 4H^+ + CH_2O \rightarrow 4FeCO_3 + 7H_2O + CO_2$
$\Delta G = -12.3$ kcal/mol

Sulphate reduction
$\frac{1}{2}SO_4^{2-} + CH_2O + \frac{1}{2}H^+ \rightarrow \frac{1}{2}HS^- + CO_2 + H_2O$
$\Delta G = -5.9$ kcal/mol

Methane formation
$CH_2O \rightarrow \frac{1}{2}CH_4 + \frac{1}{2}CO_2$
$\Delta G = -5.6$ kcal/mol

CH_2O is a simplified representation of organic matter

result in phosphorus mobilisation (Sundby, Gobeil, Silverberg & Mucci, 1992). These reactions involving the important nutrients, nitrate and phosphorus, illustrate the ways in which intertidal sediment processes can influence the productivity of the intertidal ecosystems and potentially the adjacent coastal waters. These potential effects provide a rationale for flux studies such as those of Carpenter (this volume) and similar studies in mangals (e.g. Rivera-Monroy *et al.*, 1995). Once the reservoirs of available oxidised iron and manganese are depleted, sulphate reduction begins as discussed by Ruddy (this volume).

In deep ocean waters, organic carbon inputs, microbial activity, sedimentation and bioturbation rates are relatively low and the various TEA reaction zones occur as distinct vertical layers. As the chapter by Ruddy emphasises, this is not the case in these intertidal sediments where organic inputs are high and heterogenous, giving rise to high benthic bioturbation rates. Heterogeneity may be particularly important in mangal forests where there is evidence that the different physiology and biochemistry of tree species may result in markedly different patterns of accumulation and degradation of organic matter in sediments (Lacerda, Ittekkot & Patchineelam, 1995). Together these factors mean that the different sediment reaction zones will be concentrated around micro-environments, yielding complex sediment profiles of chemicals involved in these redox cycles and a variety of resultant diagnetic mineral assemblages (Coleman, 1985). Pye *et al.* in this book consider the mineralogical evidence for the mechanisms of carbon oxidation and further emphasise that the classic theories of clear transitions from one TEA zone to another with depth in the core are inappropriate. This chapter also emphasises that microbiological investigations can provide a useful insight into the controls on these processes.

Estimations of fluxes out of these intertidal sediment systems cannot be based solely on passive and diffusional flux measurement. Rather such flux estimates must take account of the active exchange arising from burrow irrigation, the effect of the tidal 'pumping' (arising from the drainage of sediment pore waters at low tide and their recharge at high tide which can affect both water and gas exchange, Nuttle & Hemond, 1988) and the effect of uptake by algal and nutrient communities at the surface of the sediments. Many of these biologically mediated processes (bioturbation, plant growth, benthic algae growth) will have a marked seasonality in temperate regions. In addition, redox reactions in sediments can show a seasonality both directly via the seasonal input of organic matter from the growth cycle of marsh plants (at least in temperate regions) and also via the seasonal production of organic ligands in pore waters by higher plants which can regulate the sedimentary iron cycle (Luther *et al.*, 1992). Redox and pH conditions can also change on short time scales (hours) as a result of tidal pumping and intrusion of air during the tidal drying of tidal flats (Kerner & Wallmann, 1992). All these complications mean that studies of fluxes from these intertidal systems tend to take one of two forms; either involving studies of individual sediment cores to assess rates of individual processes or alternatively fluxes from the whole system. Both are difficult and uncertain as the chapters by Carpenter and Malcolm illustrate. This means that we are a long way from currently being able to effectively quantify fluxes in and out of these intertidal sediment systems. The chapter by Malcom & Sivyer in particular does, however, illustrate the potential importance of these fluxes to coastal ecosystems in general.

A further consequence of the oxidation of organic carbon by a sequence of TEAs is the production of various gaseous products including reduced sulphur species such as dimethyl sulphide (Steudler & Peterson, 1984) and carbonyl sulphide (Chin & Davis, 1993), methane and N_2O. These gases all have potential climate modifying roles (Houghton, Callander and Vaney, 1992). For the sulphur gases, saltmarshes (and possibly all intertidal systems) appear to produce relatively high fluxes per unit area, though their contributions to global fluxes, in comparison to ocean and terrestrial soil sources which occur over much larger areas, is necessarily modest (Charlson *et al.*, 1987). Methane emissions from wetlands are also potentially important (Dacey, Drake & Klug, 1994) though saltmarshes are likely to be small sources compared to terrestrial wetlands because of the greater role in marine systems of sulphate reduction which precedes methanogenesis (Table 1.1; Harriss *et al.*, 1988). However, Barber, Burke & Sackett (1988) have reported very high methane fluxes from Florida mangrove environments which they suggest may reflect rapid consumption of sulphate in these organic rich environments. Malcolm & Sivyer (this volume) indicate the potential for large scale N_2O emissions from intertidal sediments as a biproduct of nitrate reduction.

Since all these gas emissions are byproducts of organic matter decomposition they are natural sources, but the magnitude of these fluxes may be altered by human intervention both directly, via destruction of intertidal sediments, and also indirectly, via increases in atmospheric CO_2 levels (Dacey, Drake & Klug, 1994) or changes in nitrate fluxes (Malcolm, this volume). Thus while intertidal sediments are very sensitive to climate change, they are not wholly passive since they can themselves influence climate change via atmospheric trace gas cycling. However, the global significance of this is inevitably limited by their areal extent at the land/sea interface.

On a global scale, shelf sediments appear to be major sinks for organic carbon storage (Wollast, 1991) and hence important regulators of atmospheric CO_2. Within the shelf system, wetlands may contain a third of this carbon storage (Twilley, Chen & Hargis, 1992). The bulk of this wetland storage is in mangal systems, largely because of the slower decay of the woody material (Twilley, Chen & Hargis, 1992). The relatively slow degradation of the woody mangrove material, compared to the detritus of saltmarshes, also means that the residence time of carbon in the mangal system will be longer. A further consideration in some areas is the long term fate of these organic rich sediments during sea level rise. Some may be buried but others eroded and perhaps oxidised.

The other major unifying geochemical theme in intertidal systems is the adsorption of a wide range of chemicals to the sediments. As noted earlier this is a strong function of sediment particle size, but it also depends on the redox state of the sediments, the chemical behaviour of the particular component of interest, the physico-chemical nature of the sediment, and interstitial water salinity as discussed in the chapters by McDonald & Jones, Zhu & Rowland, Turner & Tyler and Rae. This adsorption allows intertidal sediments to provide an historical record of environmental contamination, provided this record is interpreted with care to allow for the complications noted above, as shown in the chapter by Rae. Intertidal sediments can represent important sinks for contaminants, carbon and nutrients, and, as with nitrate reduction (Malcolm & Sivyer, this volume), can represent a mechanism for mitigating coastal waters contamination. However, this storage can also leave these systems potentially vulnerable to pollution, though apart from the direct effects of oil pollution and enhanced algal growth due to inputs, documented pollution of intertidal systems is limited (Adam, 1990).

The other critical issue with storage of contaminants in intertidal systems is their potential for re-release – or 'chemical bombs' in the terminology of Liss *et al.* (1991). Such release could occur via erosion or changes in the chemical environment in terms of, for example, redox (Kerner & Wallmann, 1992) or salinity. An interesting natural example of such release in an undisturbed system is provided by barium and ^{226}Ra in the Ganges/Brahmaputra system. Here barium and ^{226}Ra rich sedimentary material is deposited under low salinity conditions in mangal swamps during winter high river flow conditions. In summer under low river flows, saline waters intrude further into the mangal system leading to barium and ^{226}Ra desorption and flux to sea (Carroll *et al.*, 1993). This process may be of widespread significance for barium cycling (Coffey *et al.*, in press). In the case of long lived radionuclides, the potential hazard arising from re-release from sediments persists for hundreds, or even thousands of years, but for other contaminants, cessation of inputs, dilution by uncontaminated sediments and slow degradation of 'persistent' organic molecules probably reduce the time scales over which re-release represents a major concern to tens or hundreds of years. Given human use of intertidal sedimentary environments for recreation and food supplies, adjacent waste discharge is clearly unsatisfactory.

However, the bigger threats to these intertidal systems almost certainly come from our apparently endless appetite for destroying them to reclaim agricultural land or for coastal 'development'. Faced with the inevitability of sea level rise it is to be hoped that this trend will now reverse and allow us to celebrate and encourage these beautiful and complex environments.

Acknowledgement

We would like to thank all the authors of other chapters in the book whose useful comment improved the manuscript. TDJ wishes to thank Kate Carpenter, Tom Church and Joe Scudlark for all they have taught him about saltmarshes. This chapter represents a contribution of the JONUS programme funded by MAFF, DoE and NRA. It is University of Reading PRIS Contribution No. 442.

References

Adam, P. (1990) *Saltmarsh Ecology*, Cambridge University Press, Cambridge, 461 pp.

Allen, J.R.L. & Pye, K. (1992) Coastal saltmarshes: their nature and importance. In *Saltmarshes*, 1–18, eds. J.R.L. Allen and K. Pye, Cambridge University Press, Cambridge, 184 pp.

Andrews, J.E., Brimblecombe, P., Jickells, T.D. & Liss, P.S. (1995) *An Introduction to Environmental Chemistry*, Blackwell Science, Oxford.

Barber, T.R., Burke, R.A. & Sackett, W.M. (1988) Diffuse flux of methane from warm wetlands. *Global Biogeochem. Cycl.* **2**, 411–25.

Blasco, F. (1977) Outlines of ecology, botany and forestry of the mangals of the Indian subcontinent. In *Wet Coastal Ecosystems*, 241–60, ed. V.J. Chapman, Elsevier, Amsterdam, 428 pp.

Boorman, L.A., Goss-Custard, J.D. & S. McGrorty (1989) *Climate change, rising sea level and the British Coast*, ITE Research Publication no. 1, NERC, HMSO, London, 24 pp.

Boto, K.G. and Robertson, A.I. (1990) The relationship between nitrogen fixation and tidal exports of nitrogen in a tropical mangrove system. *Est. Coastal Shelf Sci.* **31**, 531–40.

Brampton, A.H. (1992) Engineering significance of British saltmarshes. In *Saltmarshes,* 115–22, eds. J.R.L. Allen and K. Pye, Cambridge University Press, Cambridge, 184 pp.

Cadée, G.C. & Hegeman, J. (1974) Primary production of the benthic microflora living on tidal flats in the Dutch Wadden Sea. *Neth. J. Sea Res.* **8**, 260–91.

Capone, D.G. & Carpenter, E.J. (1982) Nitrogen fixation in the marine environment. *Science* **217**, 1140–2.

Carroll, J., Falkner, K.K., Brown, E.T. & Moore, W.S. (1993) The role of the Ganges–Brahmaputra mixing zone in supplying Ba and ^{226}Ra to the Bay of Bengal. *Geochim. Cosmochim. Acta* **57**, 2981–990.

Chapman, V.J. ed. (1977) *Wet Coastal Ecosystems.* Elsevier, Amsterdam, 428 pp.

Charlson, R.J., Lovelock, J.E., Andreae, M.O. & Warren, S.G. (1987) Oceanic phytoplankton, atmospheric sulphur, cloud albedo and climate. *Nature* **326**, 655–61.

Chin, M. & Davis D.D. (1993) Global sources and sinks for OCS and CS_2 and their distributions. *Global Biogeochem. Cycl.* **7**, 321–37.

Coffey, M., Dehairs, F., Collette, O., Luther, G., Church, T. & Jickells, T.D., The behaviour of dissolved barium in estuaries. *Est. Coastal Shelf Sci.* (in press).

Coleman, M.L. (1985) Geochemistry of diagenetic non-silicate minerals: kinetic considerations. *Phil. Trans. R. Soc. London* A, **315**, 39–56.

Dacey, J.W.H., Drake, B.G. & Klug, M.J. (1994) Stimulation of methane emission by carbon dioxide enrichment of marsh vegetation. *Nature* **370**, 47–9.

Doodey, J.P. (1992) The conservation of British saltmarshes. In *Saltmarshes,* 80–114, eds. J.R.L. Allen and K. Pye, Cambridge University Press, Cambridge, 184 pp.

French, J.R. & Spencer, T. (1993) Dynamics of sedimentation in a tide-dominated back barrier salt marsh, Norfolk, UK. *Mar. Geol.* **110**, 315–31.

Funnell, B.M. & Pearson, I. (1989) Holocene sedimentation on the North Norfolk barrier coast in relation to relative sea-level change. *J. Quat. Sci.* **4**, 25–36.

Halim, V. (1991) The impact of human alterations on the hydrological cycle on ocean margins. In *Ocean Margin Processes in Global Change,* 301–27, eds. R.F.C. Mantoura, J-M. Martin and R. Wollast, Wiley, Chichester, 469 pp.

Harriss, R.C., Sebacher, D.I., Bartlett, K.B., Bartlett, D.S. & Gill, P.M. (1988) Sources of atmospheric methane in the south Florida environment. *Global Biogeochem. Cycl.* **2**, 231–43.

Hinrichsen, D. (1994) Coasts under pressure. *People and the Planet* **4**, 6–9.

Houghton, J.T., Callander, B.A. & Vaney, S.K. (1992) *Climate Change 1992.* The Supplementary Report to the IPCC Scientific Assessment. Cambridge University Press, Cambridge, 200 pp.

Howarth, R.W., Marino, R., Lane, J. & Cole, J.J. (1988) Nitrogen fixation in freshwater, estuarine and marine ecosystems. 1. Rates and importance. *Limnol. Oceanogr.* **33**, 669–87.

Jelgersma, S. (1979) Sea-level changes in the North Sea basin in The Quaternary History of the North Sea. Acta Univ. Ups. Symp. *Univ. Ups. Annum. Quingentesium Celebrantis*: 2, Uppsala, 233–48.

Kerner, M. & Wallmann, K. (1992) Remobilisation events involving Cd and Zn from intertidal flat sediments in the Elbe estuary during the tidal cycle. *Est. Coastal Shelf Sci.* **35**, 371–93.

Lacerda, L.D., Ittekkot, V. & Patchineelam, S.R. (1995) Biogeochemistry of mangrove soil organic matter: a comparison between *Rhizophora* and *Avicennia* in south-eastern Brazil. *Est. Coastal Shelf Sci.* **40**, 713–20.

Liss, P.S., Billen, G., Duce, R.A., Gordeev, V.V., Martin, J-M., McCave, I.N., Meincke, J., Milliman, J.D., Sicre, M-A., Spitzy, A. & Windom, H.L. (1991) What regulates boundary fluxes at ocean margins. In *Ocean Margin Processes in Global Change*, 111–26, eds. R.F.C. Mantoura, J-M. Martin and R. Wollast, Wiley, Chichester, 469 pp.

Lugo, A.E., Brown, S. & Bronson, M.M. (1988) Forested wetlands in freshwater and salt-water environments. *Limnol. Oceanogr.* **33**, 894–909.

Luther, G.W., Kostka, J.E., Church, T.M., Sulzberger, B. & Stumm, W. (1992) Seasonal iron cycling in the salt-marsh sedimentary environment: the importance of ligand complexes with Fe(II) and Fe(III) in the dissolution of Fe(III) minerals and pyrite respectively. *Mar. Chem.* **40**, 81–103.

Mann, K.H. (1982) *Ecology of Coastal Waters*, Blackwell, Oxford, 322 pp.

Mathieson, A.C. & Nienhuis, P.H. (1991) *Intertidal and littoral ecosystems*, Elsevier, Amsterdam, 564 pp.

North Sea Task Force (1993) *North Sea Quality Status Report*. Oslo and Paris Commissions, London. Olsen and Olsen, Fredensborg, Denmark, 132 pp.

Nummedal, D., Pilkey, O.H. & Howard, J.D. eds. (1987) *Sea-level fluctuation and coastal evolution*, Spec. Pub. Soc. Economic Palaeontologists and Mineralogists **41**, Tulsa, USA, 267 pp.

Nuttle, W.K. & Hemond, H.F. (1988) Salt marsh hydrology: implications for biogeochemical fluxes to the atmosphere and estuaries. *Global Biogeochem. Cycl.* **2**, 91–114.

Nyman, J.A., Crozier, C.R. & de Laune, R.D. (1995) Roles and patterns of hurricane sedimentation in an estuarine marsh landscape. *Est. Coastal Shelf Sci.* **40**, 665–79.

Peerbolte, E.B., Eysink, W.D. & Ruardij, P. (1991) Morphological and ecological effects of sea level rise: an evaluation for the Western Wadden Sea. In *Ocean Margin Processes in Global Change*, 329–47, eds. R.F.C. Mantoura, J-M. Martin and R. Wollast, Wiley, Chichester, 469 pp.

Pethick, J.S. (1992) Saltmarsh geomorphology. In *Saltmarshes*, 41–62, eds. J.R.L. Allen and K. Pye, Cambridge University Press, Cambridge, 184 pp.

Reed, D.J. (1990) The impact of sea-level rise on coastal salt marshes. *Prog. Phys. Geog.* **14**, 465–81.

Reimold, R.J. (1977) Mangals and saltmarshes of eastern United States. In *Wet Coastal Ecosystems*, 1–18, ed. V.J. Chapman, Elsevier, Amsterdam, 428 pp.

Rivera-Monroy, V.H., Day, J.W., Twilley, R.R., Vera-Henera, F. & Coronado-Molina, C. (1995) Flux of nitrogen and sediment in a fringe mangrove forest in teminos lagoon Mexico. *Est. Coastal Shelf Sci.* **40**, 137–60.

Steudler, P.A. & Peterson, B.J. (1984) Contribution of gaseous sulphur from salt marshes to the global sulphur cycle. *Nature* **311**, 455–7.

Sundby, B., Gobeil, C., Silverberg, N. & Mucci, A. (1992) The phosphorus cycle in coastal marine sediments. *Limnol. Oceanogr.* **37**, 1129–45.

Tooley, M.J. (1992) Recent sea-level changes. In *Saltmarshes*, 19–40, eds. J.R.L. Allen and K. Pye, Cambridge University Press, Cambridge, 184 pp.

Twilley, R.R., Chen, R.H. & Hargis, T. (1992). Carbon sinks in mangroves and their implications to carbon budget of tropical coastal ecosystems. *Water Air and Soil Pollut.* **64**, 265–88.

Walker, H.J. (1990) The Coastal Zone. In *The Earth as Transformed by Human Action*, 271–94, eds. B.L. Turner III, W.C. Clark, R.W. Kates, J.F. Richards, J.T. Matthews and W.B. Meyer, Cambridge University Press, Cambridge, 713 pp.

Wiegert, R.G., Pomeroy, L.R. & Wiebe, W.J. (1981) *The Ecology of a Salt Marsh*, Springer-Verlag, New York, 271 pp.

Wigley, T.M.L., & Raper, S.C.B. (1992) Implications for climate and sea level of revised IPCC emissions scenarios. *Nature* **357**, 293–300.

Wollast, R. (1991) The coastal organic carbon cycle: fluxes, sources and sinks. In *Ocean Margin Processes in Global Change*, 365–81, eds. R.F.C. Mantoura, J-M. Martin and R. Wollast, Wiley, Chichester, 469 pp.

Woodroffe, C.D. (1990) The impact of sea-level rise on mangrove shorelines. *Prog. Phys. Geog.* **14**, 483–520.

2

○ ○ ○ ○ ○ ○ ○ ○ ○ ○ ○ ○ ○ ○ ○ ○ ○ ○ ○

Trace metals in deposited intertidal sediments

J.E. Rae

Introduction

Trace metals, for example elements from the first row of transition metals including cobalt, nickel, copper and zinc (Co, Ni, Cu and Zn), occur naturally in intertidal sediments. The natural (or 'background' inputs) reflect a combination of the composition of drainage-basin rocks and marine-derived sediment. The global anthropogenic input of many trace metals to the environment currently equals or exceeds the amount released by weathering (de Groot, Salomons & Allersma, 1976), and the extent of anthropogenic influence in intertidal sediments is particularly high, since intertidal areas are often considered as convenient dumping grounds for industrial and other waste.

Anthropogenic inputs to intertidal environments are often direct, through point-source waste disposal, but they are also indirect, from riverine, marine and/or atmospheric sources. Trace metals are partitioned between each component of the intertidal sediment–water system: they are found in solution ('bulk' water or interstitial water) and associated with suspended and deposited sediments. This chapter is concerned with the biogeochemistry of trace metals in deposited intertidal sediments. Two main sections follow: in the first, an overview of surface sediments and sediment depth profiles is presented, and in the second, a case study is given of the historic record of Zn from saltmarsh sediments in the Severn Estuary, UK.

2.1 Overview of trace metals in deposited intertidal sediments

2.1.1 Surface sediments

Concentrations and partitioning of trace metals and relation to bioavailability

Concentrations of trace metals (particularly the pollutants, Cr, Mn, Fe, Co, Cu, Ni, Cd, Hg and Pb) are commonly reported in surficial intertidal surface sediments, for example in the St Lawrence Estuary, Canada (Loring, 1978), San Francisco Bay, USA (Thompson-Becker & Luoma, 1985), the Kelang Estuary, Malaysia (Law & Singh, 1988), the Brisbane River Estuary, South East Queensland, Australia (Mackey, Hodgkinson & Nardella, 1992), the Severn Estuary, UK (French, 1993a, b), and the Sado Estuary, Portugal (Cortesäo & Vale, 1995). Natural trace metal concentrations are less commonly reported, although some useful compilations of data exist, e.g. Windom *et al.* (1989) have considered background data for estuarine and coastal sediments of the southeastern United States.

Estuarine environments in particular are often sites of intense human and animal activity (e.g. sites of leisure pursuits and breeding grounds for many species of birds) so the level of contamination of intertidal sediments is of particular interest in relation to environmental health. Total concentrations of individual trace elements in UK estuaries, for example, vary widely (Table 2.1), reflecting the natural sediment characteristics (e.g. organic content and surface area), and the level of anthropogenic contamination of individual systems (Bryan & Langston, 1992). The order of variability of concentrations of individual elements in Table 2.1 is Sn > As > Cu > Pb > Hg > Ag > Zn > Cd > Se > Cr > Mn > Co > Ni > Fe, which to some extent can be considered as the order of anthropogenic influence.

Trace metals are held in deposited sediments in a number of different 'sites'. First of all there are metals in the mineral lattices (often referred to as the 'residual' phase metals). In uncontaminated sediments, this is normally the most important site quantitatively. Secondly, trace metals, particularly anthropogenically introduced metals, occur associated with particle surfaces, hence there is often a significant correlation between trace metal content and grain-size or surface area, which is in turn related to the sediment mineralogy (Figure 2.1). There has been a lot of detailed work in recent years on the mode of occurrence of trace metals in sediments and one of three different approaches is normally taken.

Table 2.1 Average concentrations of metals ($\mu g g^{-1}$ dry wt) in sediments from 19 UK estuaries[a] (placed in order of decreasing Cu concentration) (after Bryan & Langston, 1992)

Site	Ag	As	Cd	Co	Cr	Cu	Fe	Hg	Mn	Ni	Pb	Se	Sn[b]	Zn
Restronguet Creek	3.76	**1740**	1.53	21	32	**2398**	**49071**	0.46	485	**58**	341	—	55.9	**2821**
Fal	1.37	56	0.78	9	28	648	28063	0.20	272	23	150	—	39.5	750
Tamar	1.22	93	0.96	21	47	330	35124	0.83	590	44	235	1.40	8.3	452
Gannel	**4.13**	174	1.35	**26**	24	150	25420	0.08	649	38	**2753**	—	8.5	940
Tyne	1.55	24.8	**2.17**	11	46	92	28206	0.92	395	34	187	1.23	5.4	421
Mersey	0.70	41.6	1.15	13	84	84	27326	3.01	**1169**	29	124	0.30	8.3	379
Humber	0.42	49.8	0.48	16	77	54	35203	0.55	1015	39	113	0.84	5.05	252
Medway	1.45	18.4	1.08	11	53	55	32216	1.00	418	26	86	0.48	3.4	220
Poole	0.82	14.1	1.85	11	49	50	29290	0.81	185	26	96	**1.51**	7.4	165
Severn	0.42	8.9	0.63	15	55	38	28348	0.51	686	33	89	0.23	8.0	259
Hamble	0.16	18.4	0.34	10	37	31	28132	0.43	241	19	56	0.41	3.9	105
Loughor	0.18	17.5	0.47	10	**207**	27	19337	0.16	597	21	48	—	**161**	146
Dyfi	0.19	9.7	0.62	17	32	24	45683	0.12	1127	33	166	0.20	0.98	212
Wyre	0.21	6.4	0.35	8	37	20	16970	1.52	590	17	44	—	3.15	122
Avon	0.06	13.0	0.08	10	28	18	18361	0.12	326	23	68	—	3.9	82
Teifi	0.09	10.6	0.17	10	29	13	30280	0.04	684	23	25	0.16	1.14	87
Axe	0.13	4.8	0.17	7	27	12	14004	0.20	248	14	26	—	1.39	76
Rother	0.17	12.4	0.13	6	29	11	15648	0.09	259	15	20	0.18	0.62	46
Solway	0.07	6.4	0.23	6	30	7	14816	0.03	577	17	25	0.11	0.40	59

[a] Data are for the < 100 μm fraction of surface sediment following digestion in HNO_3.
[b] A large proportion of Sn in some samples occurs as cassiterite and is not dissolved in HNO_3 (fusion technique necessary).
Figures in bold are the highest concentrations.

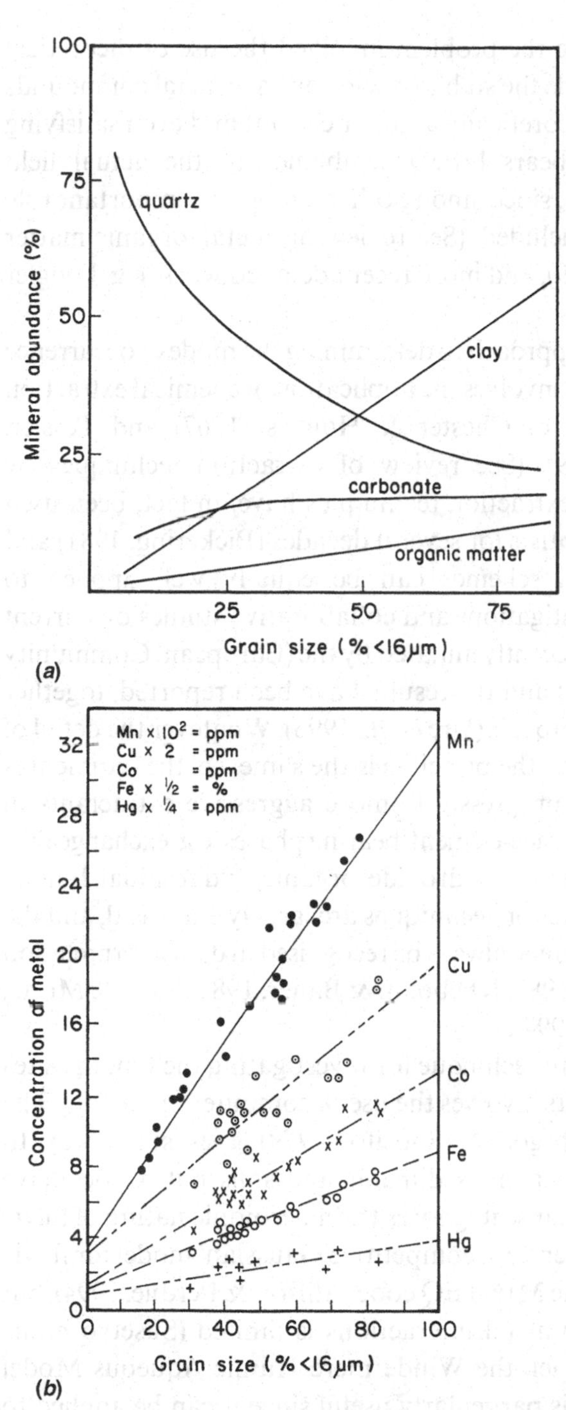

Figure 2.1 (*a*) Relationship between mineralogical composition and grain-size distribution (% < 16 μm) for samples from the same locality (after de Groot, 1976). (*b*) Relationship between trace element concentrations and grain-size distribution (% < 16 μm) for sediment samples (after de Groot, 1976).

An early way of tackling the problem involved the use of theoretical Eh–pH diagrams to examine the stability fields of trace metal compounds (e.g. Hem, 1972). This is theoretically a rather elegant and even satisfying approach, but probably bears little resemblance to the actual field distribution of trace metals, since among other factors the important role of organic matter is not included. (See review on metal–organic matter interactions by Saxby (1969), and more recent detailed work, e.g. Fengler *et al.*, 1994.)

A second and well-used approach to determining the mode of occurrence of trace metals in sediments involves the application of chemical extraction techniques such as those of Chester & Hughes (1967) and Tessier, Campbell & Bisson (1979). (See review of extraction techniques by Towner, 1984.) Chemical extraction techniques have, in fact, been used and developed by soil scientists for several decades (Pickering, 1981) and many of these extraction schemes can be equally well applied to sediments. A series of investigations and collaborative studies on current extraction techniques was recently initiated by the (European) Community Bureau of Reference (BCR) and the results have been reported, together with agreed procedural protocols (Ure *et al.*, 1993). Whatever the detail of any one technique, however, the principle is the same, i.e. the sediment is sequentially leached with progressively more aggressive extractants in order to define a number of trace-element-bearing phases, e.g. exchangeable, carbonate, Fe and Mn oxide and hydroxide, organic, and residual. It must be noted that all such extraction techniques are largely empirical, and the limitations of the approach must always be recognised in data interpretation (Martin, Nirel & Thomas, 1987; Kheboian & Bauer, 1987; Nirel & Morel, 1990; Whalley & Grant, 1994).

The third and most recent technique for investigating the binding sites for trace metals in sediments involves the use of computer modelling. The model GEOCHEM (Mattigod & Sposito, 1979) goes some way to establishing speciation in waters, sediments and soils but it does have shortcomings, not least because it ignores the macroionic nature of fulvic and humic acids. More recently a 'competitive Gaussian' model for fulvic acid was introduced into the MINTEQ code (Allison & Perdue, 1994), but the available database for metal interactions is limited (Susetyo *et al.*, 1991). Another recent model, the Windermere Humic Aqueous Model (WHAM) (Tipping, 1994) is particularly useful since it can be applied to equilibrium speciation problems involving waters, sediments and soils with humic substances present in dissolved and particulate form. WHAM is a combination of several submodels: the Humic Ion-Binding Model V

(Tipping & Hurley, 1992; Tipping, 1993a, b), and models of inorganic solution chemistry, precipitation of aluminium and iron oxyhydroxides, cation exchange on a representative clay and the adsorption–desorption reactions of fulvic acids (Tipping & Woof, 1990, 1991). Other recent models include SCAMP (Surface Chemistry Assemblage Model for Particles) which is now being developed as part of a large project in the UK (Land–Ocean Interaction Study (LOIS)). Computer models, although an important advance, should not be treated as a panacea. They are, for example, currently limited in the number of variables which can be considered. Perhaps we should view these models as an adumbration of detailed computer models of the future, which may more closely resemble the real situation.

An important reason for studying the concentrations and partitioning of trace metals in intertidal sediments is to enhance our understanding of bioaccumulation and biological effects of trace metals in intertidal environments. Deposited sediments are important for at least three reasons: firstly, because uptake of metals by organisms from solution is very significant and concentrations of metals in interstitial and overlying water can be controlled by equilibria between dissolved metals and those adsorbed by sediment particles; secondly, concentrations of most trace metals in sediments are orders of magnitude higher than concentrations in solution; lastly, some trace metals, e.g. As, Sn, Hg and Pb, can be converted to more toxic organic (methyl) compounds during their association with sediment (Bryan & Langston, 1992). Filter feeders and burrowing organisms are particularly at risk from sediment-bound trace metals, but biomagnification with increasing trophic levels along food chains can also occur (e.g. Goede, 1985). An important question must, however, be addressed: that is, how accurately can we determine the biological availability (bioavailability) of sediment-bound trace metals? This question has been usefully discussed by Luoma (1989), who draws attention to pitfalls awaiting the uncritical researcher. One approach to investigating metal bioavailability in sediments is through the use of partial chemical extractions such as those previously discussed (e.g. Tessier & Campbell, 1987; Tessier *et al.*, 1984), but Luoma (1989) lists a number of inherent limitations:

(1) since extraction techniques rarely (if ever) remove metals from entirely specific components of the sediment, correlations of biological availability with specific metals components is problematic;
(2) reactions in solution (such as complexation), and competition

among cations for transport into the animal are not accounted for;
(3) biological uptake processes are complex; for example, digestion is a 'flexible, adaptive, multi-faceted living process that can change in response to environmental conditions or with life history'.

Despite such limitations, significant relationships have been observed between concentrations of metals in deposit-feeding organisms and in the sediments (Bryan & Langston, 1992), Table 2.2. Some of the better relationships that have been found depend on the use of metal/Fe or metal/% organic matter ratios in the sediments. Luoma & Bryan (1978), for example, in a study of English estuaries, showed that 80% of the variance in Pb concentrations in *Scrobicularia plana* could be explained by the ratio of Pb/Fe, (the weak acid (IN HCl) extractable fractions) in the sediments, and Langston (1982) demonstrated decreased bioavailability of total Hg where sediments are rich in organic matter.

A number of future research needs in this area are suggested by Bryan & Langston (1992), particularly:

(1) the mechanisms by which metal-binding sediment components (e.g. oxides of Fe and Mn and organics) control bioavailability in the field must be investigated (It is uncertain whether the effects of such components are important owing to their control over the levels of dissolved metals, or because they control the uptake of

Table 2.2 Examples of significant relationships between metal concentrations in benthic species and surface sediments from a range of estuaries (after Bryan & Langston, 1992)

Organism	Metal	Sediment extract best predicting bioavailability	Reference
Macoma balthica	As	As/Fe in 1N HCl ($r = 0.94$)	Langston (1986)
(bivalve)	Ag	Ag in 1N HCl ($r = 0.83$)	Bryan (1985)
	Hg	Hg in HNO_3 extract/% organics ($r = 0.83$)	Langston (1982)
	Pb	Pb/Fe in 1N HCl ($r = 0.76$)	Bryan (1985)
Nereis diversicolor	As	As/Fe in 1N HCl ($r = 0.74$)	Langston (1980)
(polychaete)	Ag	Ag in 1N HCl ($r = 0.79$)	Bryan (1985)
	Hg	Hg in HNO_3 extract/% organics ($r = 0.61$)	Langston (1986)
	Pb	Pb in HNO_3 extracts ($r = 0.73$)	Luoma & Bryan (1982)

metals from ingested sediment in the gut.);

(2) more work is required on the composition of interstitial waters in relation to bioavailability and toxicity;

(3) experimental work on pH and metal/metal interactions in the digestive tract is required for a range of sediment-ingesting species.

Methods used for reporting trace metal concentrations of sediments

It is well known that the trace metal content of a sediment is often to a large extent a function of its chemical and mineralogical characteristics. It is, therefore, very important to use a reliable 'normalising' technique for reporting trace metal concentrations. The key sediment characteristic is surface area or particle size since many of the trace metal binding components (e.g. organic matter, Fe and Mn oxides and hydroxides) are very well correlated with both characteristics.

Methods for correcting for grain-size effects in studies on heavy metal concentrations in estuarine and coastal sediments have been discussed by Ackermann (1980). There is, unfortunately, no one standard method for particle-size normalisation and a wide range of techniques are in use (Table 2.3). The method which often involves the least effort is the correction which uses comparison with rubidium (Rb) as a conservative element (Ackermann, 1980). This technique relies on the fact that Rb has a similar ionic radius to potassium (K) and so substitution of Rb for K will take place in clay minerals. Furthermore, Rb is present in the sand fraction in very much smaller concentrations than in the clay or silt fraction and concentrations of the element in sediments are rarely influenced by anthropogenic activity. Another advantage of the use of Rb is that it is often routinely analysed by X-ray fluorescence along with a suite of pollutant trace metals.

Variability of trace-metal concentrations of surface sediments with space and time and implications for field sampling strategies

Any consideration of environmental quality issues based on concentrations of trace metals in intertidal sediments must also address the question of variability of these concentrations in space and time. Spatial variation of trace metal concentrations in an estuarine system clearly depends on the number and magnitude of point-source inputs and on the degree of sediment mixing. In a physically well-mixed system, the degree of spatial variation might be expected to be slight. Allen (1987c) and French (1993a) have investigated the spatial distribution of trace metals (and

Table 2.3 Methods for the reduction of grain-size effects in sediment samples (references listed are examples of the applications and/or discussion of the use of the techniques)

Method	Example	Reference
1. Trace metal analysis of a specific grain size fraction of sediment	< 64 μm (sieving) < 20 μm (sieving) < 2 μm (settling tube)	O'Reilly Wiese, Bubb & Lester (1995) Ackermann, Bergmann & Schleichert (1983) Banat, Forstner & Muller (1972)
2. Correction for inert mineral constituent	Quartz correction i.e. trace metal (qtz corrected) $= \frac{\text{trace metal (observed)} \times 100}{(100 - \text{quartz }\%)}$	Thomas & Jaquet (1976)
3. Ratios of trace metals to 'conservative' elements	$\frac{\text{trace metal}}{\text{Al}}$	Cortesäo & Vale (1995) Din (1992)
4. Use of regression curve based on	Grain size e.g. trace metal concentrations at 50% < 16 μm	Salomons & Mook (1977)
	Conservative elements e.g. Cs, Rb, Li, Al	Ackermann (1980) Allen & Rae (1986) Loring (1990) Rowlatt & Lovell (1994) Oliver (1973)
	Surface area	

other components) in contemporary muddy sediments from the Severn Estuary, UK, a large and relatively well-mixed system. A streamwise variation in grain-size corrected trace metal values was observed (Figure 2.2*a*) and was interpreted as reflecting a combination of inputs from land-based sources and re-working of pollutant-rich saltmarsh sediments. A similar study for the Rhine–Meuse Estuary System (de Groot *et al.*,

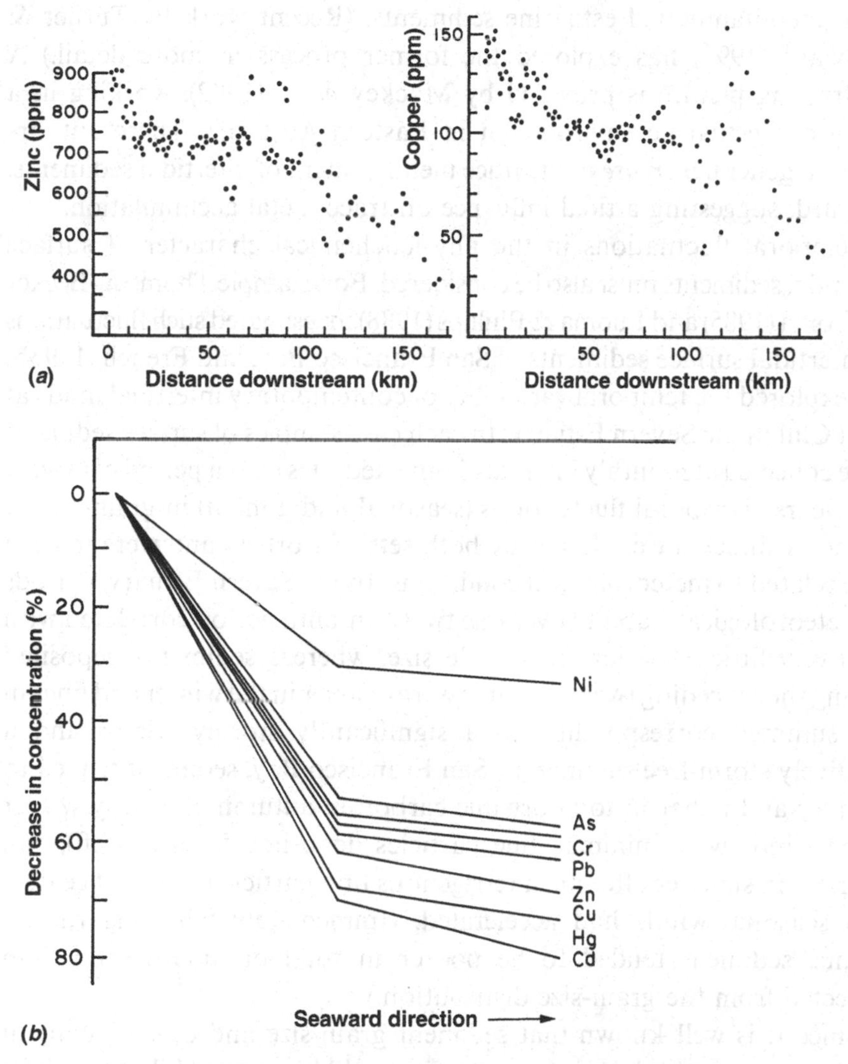

Figure 2.2 (*a*) Spatial (downstream) variation in grain-size corrected concentrations of zinc and copper, in surface sediments from the Seven Estuary, UK (after French, 1993a). [Note: data are presented in smoothed form calculated using a three point moving average.] (*b*) Spatial (downstream) variation in trace metal concentrations of sediments from the Rhine Estuary, expressed as percentages of the contents in fluvial sediments (after de Groot, 1976).

1976) (Fig. 2.2*b*) also demonstrated a general decrease in sediment trace metal contents in a seaward direction, although grain-size variation does not appear to have been taken into account. De Groot *et al.* (1976) suggest that the downstream trend in the Rhine–Meuse system is a function of a combination of two very different processes: chemical desorption of trace metals as the ionic strength of estuarine waters increases downstream, and physical mixing of polluted fluvatile sediments with uncontaminated estuarine sediments. (Recent work by Turner & Millward (1993) has explored the former process in more detail.) A contrasting picture is provided by Mackey *et al.* (1992), working in a mangrove estuarine environment in Eastern Australia. These authors report a general *increase* in the trace metal content of intertidal sediments, seaward, suggesting a tidal influence on trace metal accumulation.

Temporal fluctuations in the physicochemical character of surface intertidal sediments must also be considered. For example Thomson-Becker & Luoma (1985) and Luoma & Phillips (1988) investigated such fluctuations in intertidal surface sediments of San Francisco Bay, and French (1993b) has explored the temporal variability of contemporary intertidal muds at Aust Cliff in the Severn Estuary. In each case samples of surface sediment were collected at monthly intervals from fixed sites over a period of at least two years. Temporal fluctuations (seasonal and annual) in grain-size of surface sediments were found by both sets of workers and were in each case related to meteorological conditions. In the Severn Estuary, periods of meteorological stability over one twelve-month period correlated with relatively little variation in particle size, whereas sediments deposited during the preceding twelve months were coarser in the winter and finer in the summer, corresponding to a significantly stormy winter and a relatively storm-free summer. In San Francisco Bay, sediment tended to be finer and richer in total organic carbon in autumn and early winter when winds were minimal (fine particles deposited from run-off), and coarsest in summer after most terrigenous fine-particle input had ceased, and seasonal winds had accelerated. (Immediately following run-off events, sediment tended to be poorer in total organic carbon than expected from the grain-size distribution.)

Since it is well known that sediment grain-size and organic content influence trace metal concentrations, it would be expected that temporal variability in trace metal concentrations would also be found. This is indeed the case (French, 1993a), but additional controls appear to also exert an influence: zinc exhibits a temporal variability even after grain-size correction in Severn Estuary sediments (French, 1993b), probably reflecting a number of processes including variability of anthropogenic inputs and of riverine discharge rates. Araujo, Bernard & Van Erieken (1988), working in the Scheldt Estuary, sampled sediment at two stations periodically during two years. The standardised values of the composition of the bulk sediments and of the fraction < 63 μm showed large variations in trace metal content, with the highest values found in samples collected in June and/or September. It is suggested that these high values may be more closely related to the increase in bacterial activity during the summer, after the spring phytoplankton bloom, than they are to an increase in anthropogenic input.

The design of a field sampling strategy for investigating temporal variability in heavy metal concentrations of surface sediments clearly needs careful thought applied to it. There are two potential problems: firstly, the problem of confounding of temporal differences by spatial variation, and secondly, the problem of variations at shorter time scales than the scale of interest. For example, to demonstrate differences from one season to another, it must be shown that there is more variation between seasons than is found from time to time within a season. These problems are discussed by Morrisey *et al.* (1994), in a study of the distribution of Cu, Pb and Zn in the sediments of Botany Bay. The authors suggest that nesting of spatial and temporal scales should be done in a pilot study, and scales at which significant variation is not detected can then be omitted from the subsequent sampling.

2.1.2 Sediment depth profiles as records of intertidal pollution history

Trace metals in intertidal sediment depth profiles have been extensively studied since the late 1970s. There were two key factors which resulted in the proliferation of studies at this time: firstly, a rise in environmental awareness led to interest in using sediment depth profiles as historic records of pollution, and secondly, radiometric techniques for sediment dating became well established. In particular, ^{210}Pb dating of modern sediments has been widely used since it was first outlined by Goldberg (1963) and several workers have used the technique for dating intertidal sediments (e.g. McCaffrey, 1977; Goldberg, *et al.*, 1979; McCaffrey & Thompson, 1980; Sharma *et al.*, 1987; Bricker, 1990 and Buckley, Smith & Winters, 1995), all of whom have investigated trace metal concentrations in sediment depth profiles.

A number of useful reviews exists on the use of sediment cores to reconstruct historical pollution trends (Alderton, 1985; Farmer, 1991;

Valette-Silver, 1992, 1993) and there are perhaps two key issues which emerge. Firstly, sediment depth profiles (including those from intertidal environments, e.g. Buckley *et al.*, 1995) often reflect the globally averaged profiles of historic sediment contamination (Figure 2.3). The global pattern is one of significant heavy metal contamination of the environment, beginning during the last decades of the nineteenth century and reaching maximum levels around 1960–70 (Valette-Silver, 1993). Concentrations of heavy metals in fine-grained intertidal sediments commonly range from 'background' levels of a few $\mu g g^{-1}$ or tens of $\mu g g^{-1}$ up to (commonly) several hundred $\mu g g^{-1}$ for polluted sediments (Valette-Silver, 1993). The second key issue to emerge from the literature is also a caution: that is, the *character of the sediments* (e.g. grain-size, mineralogy, organic carbon

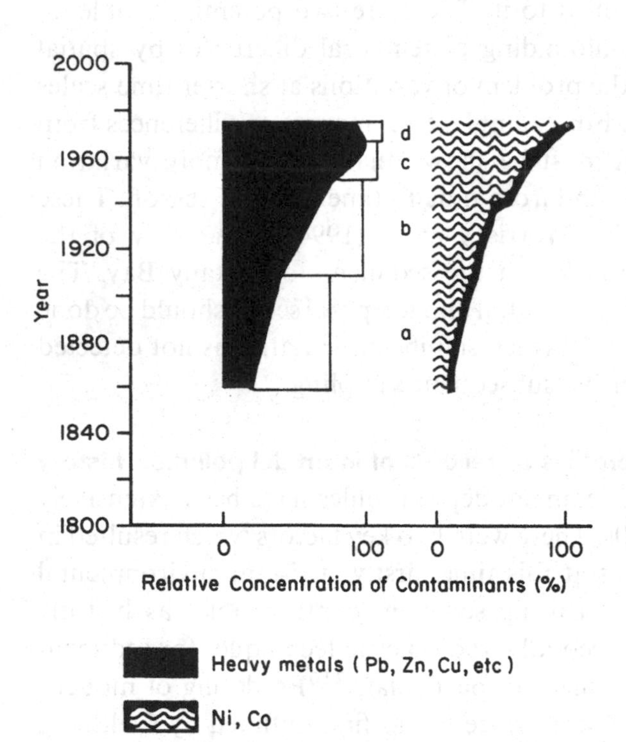

Figure 2.3 Globally averaged profiles of historic sediment contamination by trace metals, in industrialised countries, providing a general idea of the evolution of contamination versus time (years). 100% represents the maximum value reached in cored sediments and 0% the lowest value measured. These values are rough estimates, extrapolated from results reported in the literature to 1990. Section (a) represents a background or low contamination period, section (b) corresponds to the increase in sediment contamination around World War I. Section (c) visualises the rapid increases occurring after World War II that reach a maximum in the mid-1970s. Section (d) represents the recent decrease observed in concentrations of many trace metals (after Valette-Silver, 1993).

content), together with the *diagenetic history* can be important in influencing trace metal concentrations and bioavailability. In other words, it is necessary to establish, and where appropriate correct for, the sediment characteristics before confirming historic trace metal records, in addition to considering the extent of influence of diagenetic processes (Farmer, 1991; Rae & Allen, 1993; Zwolsman, Berger & Van Eck, 1993). As discussed in the previous section, pollutant trace metals are predominantly held in association with the surfaces of sedimentary materials, in a number of different 'phases' which can be experimentally defined (e.g. exchangeable, carbonate, co-precipitated with Fe and Mn oxides, organic-bound (Tessier *et al.*, 1979)). It is not surprising, therefore, to find that sediment characteristics such as grain-size and organic carbon content, and diagenetic processes such as reduction of Fe and Mn oxides, are important in influencing trace metal concentrations in sediment depth profiles (Allen, Rae & Zanin, 1990; Farmer, 1991). An illustration of the influence of sediment characteristics is given in Rae (1980), (Figure 2.4), in which an apparent historic record of mercury (Hg) pollution from an estuarine sediment core (Wyre Estuary, UK) is illustrated. The increase in sediment Hg content above 10 cm may appear to reflect a simple historic increase in environmental Hg pollution, but closer inspection reveals a

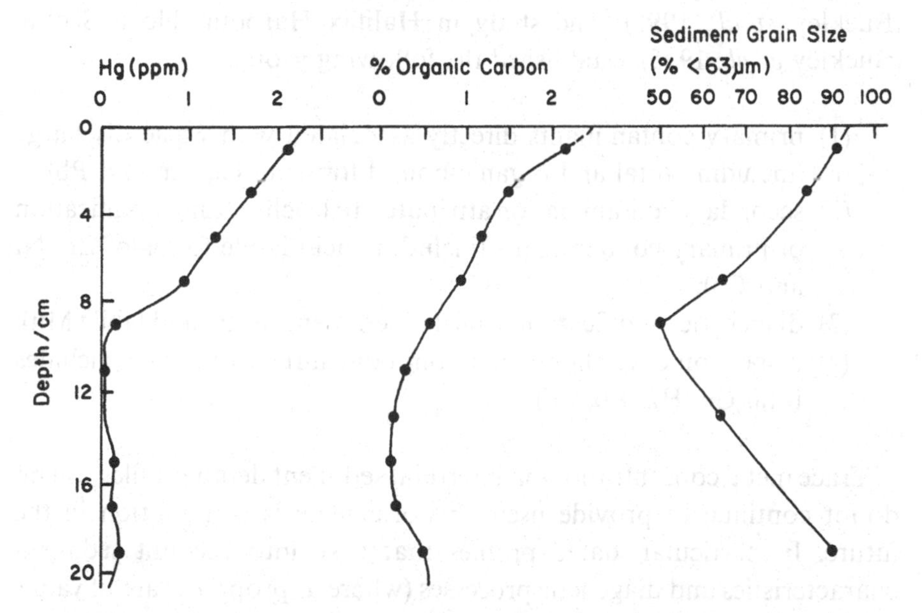

Figure 2.4 Results of the analysis (Hg, organic carbon and grain size (% < 63 μm)) for a sediment core from the Wyre Estuary, UK. Values of Hg below 10 cm represent 'background' levels with an *apparent* increase in contamination from 10 cm depth to the sediment surface (after Rae, 1980).

corresponding increase in organic carbon content and percentage of fine sediment (% < 63 μm). It is likely, therefore, that the influence of sediment characteristics is here masking the historic pollution record. (The increase in fines below 10 cm has little effect on the very low 'background' Hg concentrations.) Intertidal sediments are particularly prone to variability in sediment characteristics with depth, when compared to other sedimentary environments such as lakes, in which historic pollution trends are also frequently established. The greater variability in sediment characteristics with depth in intertidal systems results primarily from tidal and wave action which can have profound effects in influencing particle size and sorting. Diagenetic history in intertidal sediment profiles may also be more complex than in lakes, since intertidal environments may be more subject to reworking and/or more rapidly fluctuating pore-water compositions. (It should be said, however, in their favour, that intertidal environments are not subject to acidification which often complicates the use of lake sediments as historical records of pollution (Valette-Silver, 1993).)

Finally, a recent approach to the study of trace metal distribution in sediment depth profiles deserves mention. This is a factor-analysis technique which is used to determine the main environmental condition prevailing at the place and time when the sediment was deposited, or the main process responsible for modification of the sediment after deposition (Buckley *et al.*, 1995). The study in Halifax Harbour, Nova Scotia, (Buckley *et al.*, 1995), established the following groups:

(1) primary contaminants directly associated with waste discharge (including total and organic-bound forms of Cu, Zn and Pb);
(2) secondary contamination attributed to leaching and modification of primary contaminants (including acid labile forms of Zn, Ni and Cu);
(3) diagenetic modification of buried sediments (total and labile Mn);
(4) dispersion of contaminants from land surface drainage (includes total Cu, Hg, Pb, Zn).

Trace metal concentrations in intertidal sediment depth profiles will no doubt continue to provide useful historic records of pollution in the future. In particular, dated profiles that take into account sediment characteristics and diagenetic processes (where appropriate) are of value. Future studies will perhaps be more process-oriented, e.g. by combining pore-water analysis with solid phase data and through a detailed investigation of the nature and significance of the organic matter present.

Lastly, there is an urgent need for more biogeochemical work on organometallic forms of trace metals owing to their presence in intertidal sediments and their toxicity at relatively low concentrations; this work may well profit from a detailed appreciation of microbiological processes.

2.2 The record of zinc pollution in a sediment depth profile at Tites Point, Severn Estuary, UK: a case study

2.2.1 Introduction

The Severn Estuary (Figure 2.5) lies on the west coast of Britain, facing the prevailing winds and Atlantic swell. It is a large (*c.* 100 km long), macrotidal system (12.3 m mean spring tidal range at Avonmouth), swept by vigorous currents (Crickmore, 1982; Shuttler, 1982; Uncles, 1984). Urban and industrial effluent from three main areas – southeast Wales, the Bristol region, and the West Midlands – is currently being brought into the estuary (Radford, 1982; Morris, 1984; Owens, 1984).

There has been much work reported in recent years on the sedimentology and stratigraphy of the tidally influenced marshes and high tidal mud-flats (Allen, 1985a, b, 1986, 1987a, b, c, d, 1988, 1990a, b, c, d, e, f, 1992a, b; Allen & Rae, 1986, 1987, 1988).

The formation of mud-flats, which develop into saltmarshes, is occurring in response to an apparently climate-driven, estuary-wide cycle of erosion

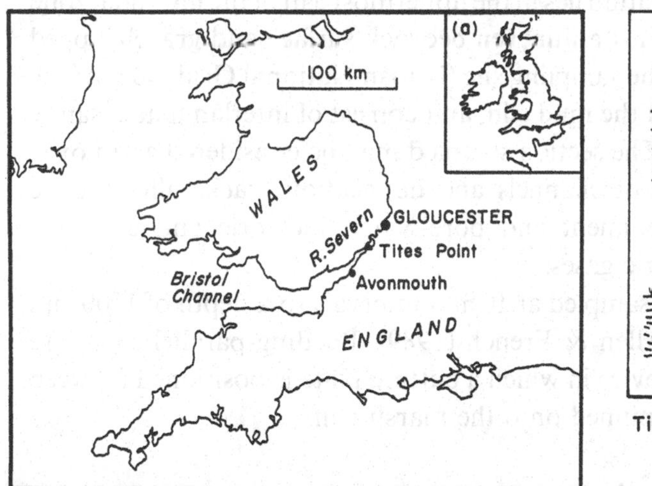

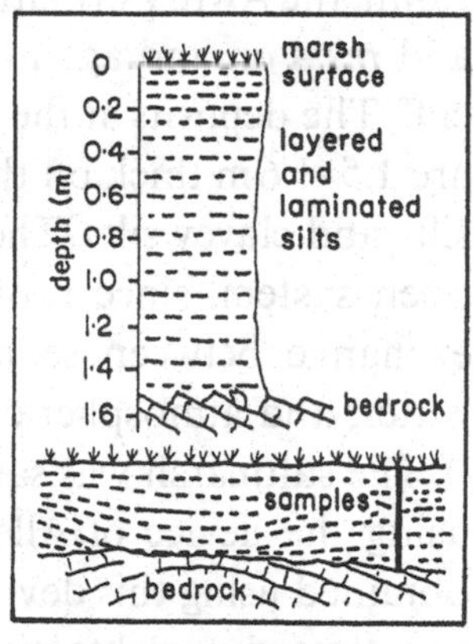

Figure 2.5 The Severn Estuary and Tites Point; setting in southwest England; a schematic vertical profile of the saltmarsh cliff (after Rae & Allen, 1993).

and accretion (Allen & Rae, 1987), with a decadal to century time scale. Mud-flat and saltmarsh deposits build up in the intertidal zone, but later become eroded vertically and horizontally (toward land) to form a wave-cut platform and bold cliff, on and against which a new mud-flat and then a marsh grows up (Allen, 1992a, b). At least three repetitions of this cycle have occurred throughout the Severn Estuary, resulting in present-day saltmarshes that are visibly terraced, descending stair-like toward the estuary, each terrace overlying an erosively based mud deposit that offlaps the preceding unit (Allen & Rae, 1987).

Early work on trace metals in sediment depth profiles in the Severn Estuary provided an overview of trace metal concentrations in marsh sediments, in order to establish a 'chemostratigraphy' (Allen, 1987c; Allen & Rae, 1987; Allen, 1988). Later work (Allen *et al.*, 1990) provided a preliminary investigation of the post-depositional behaviour of Cu, Zn and Pb at one location (Tites Point). Subsequent intensive sampling of an adjacent profile has allowed a more detailed appreciation of trace metal concentrations and behaviour.

2.2.2 Site location and sampling methods

The location of Tites Point and the sampling site are given in Figure 2.5, together with a schematic vertical profile of the saltmarsh cliff. A detailed description of the field context is given in Allen *et al.* (1990). The deposit sampled is attributed to the Awre Formation, initiated during the second depositional cycle, which began late in the nineteenth century. At Tites Point, the Awre Formation lies in the uppermost 3 m of the intertidal zone and rests on and against an uneven bedrock surface and gravel-capped cliff. The deposits at the sampling site (British National Grid SO 688046) are 1.5–1.6 m thick on the mud cliff, and consist of interlaminated sandy silts and clayey silts. The section sampled may be considered as an oxic, open system, since root channels and desiccation cracks allow active exchange between sediment and pore-water, meteoric and estuarine water, and atmospheric gases.

The saltmarsh was sampled at 10 mm intervals to a depth of 1500 mm using the device of Allen & French (1989). Bedding-parallel slices are obtained using this device in which a cutting plate is positioned between two slotted uprights pinned onto the marsh cliff.

2.2.3 Laboratory methods

Sediments were air-dried in the laboratory and lightly disaggregated. Organic carbon determinations were carried out by the method of

Gaudette *et al.* (1974), in which exothermic heating and oxidation with K_2CrO_7 and concentrated H_2SO_4 are followed by titration of the excess dichromate with 0.5N $Fe(NH_4)_2(SO_4)_2$ $6H_2O$. (This method excludes elementary carbon which in intertidal sediments is commonly present as coal, coke and fly-ash particles (Allen, 1987a, b).)

The sediment samples were sequentially leached in accordance with the method of Tessier *et al.* (1979) in which five fractions are experimentally defined. The extracts were analysed by atomic absorption (Perkin Elmer 3030), (Pb and Rb), and inductively coupled plasma spectrometry (ARL 35000), (Cu, Zn, Cd, Ni, Fe and Mn). Instrumental precision was better than 1% and accuracy was ensured by analysing a range of international reference sediments.

The sediment profile was dated using ^{210}Pb and a number of other techniques (Allen *et al.*, 1993).

2.2.4 Results and discussion

Results of dating the sediments at Tites Point are presented in Allen *et al.* (1993) and a preliminary assessment of the significance of organic matter concentrations in relation to trace metal mobility (Cu, Zn and Pb) is given in Rae & Allen (1993). A complete report of trace metal concentrations and behaviour at this site will be given in Allen & Rae (in preparation). There follows here a summary of the results for Zn, in order to demonstrate in broad terms (i) the approach to consideration of the influence of sediment characteristics and early diagenetic processes on trace metal profiles, and (ii) that the record *can* be, in some instances, straightforward.

The variation in total sediment Zn with depth is illustrated in Figure 2.6. There is a clear trend of essentially constant values of total Zn (around 150 ppm) from 1500 mm to approximately 1000 mm, with a steady increase in concentrations from 1000 mm to 225 ppm at 400 mm. Concentrations of Zn remain at approximately this latter value until they fall to 200 ppm in the uppermost 50 mm. In contrast to this clear trend, concentrations of zinc in the 'residual' phase (non-pollutant zinc) remain approximately constant with depth (~75 ppm) (Figure 2.7). We are now faced with the problem of interpretation and the key question is: does the well-defined depth trend of total Zn represent a historic pollution record or is it partly (or entirely) a function of sediment characteristics and/or early diagenetic processes?

The most important sediment characteristic is usually the sediment grain-size. The abundance of Rb in sediments is an established and reliable grain-size proxy (Allen & Rae, 1987; Allen, 1987c), increasing with clay-mineral content. There appears to be a slight but systematic increase in fines with depth below 220 mm and a decrease in fines with depth from the marsh surface to 220 mm. It is unlikely, however, that these grain-size trends are of a magnitude that would interfere with the broad picture of historic pollution at this site.

Having established that grain-size variability is not significant in this case, the next step is to investigate any perturbations in the historic pollution record owing to early diagenetic processes. One approach to this investigation is through an examination of depth trends of Zn in 'carbonate', 'Fe and Mn oxide' and 'organic' phases (Figure 2.7). (The amount of Zn at Tites Point in the exchangeable phase (generally < 1 ppm) is too low to influence the interpretation of the total zinc profile.) The question now is as follows: to what extent, if at all, are the profile of Zn in carbonate, Fe and Mn oxide and organic phases a function of early diagenetic processes? Calculation of the Zn content at each depth, in each phase, as a percentage of the 'pollutant' (i.e. non-residual) zinc at corresponding depth intervals provides a clue, although the result of sequential extraction techniques should always be used with caution and with supporting data where possible (see Section 2.1.1). In this case,

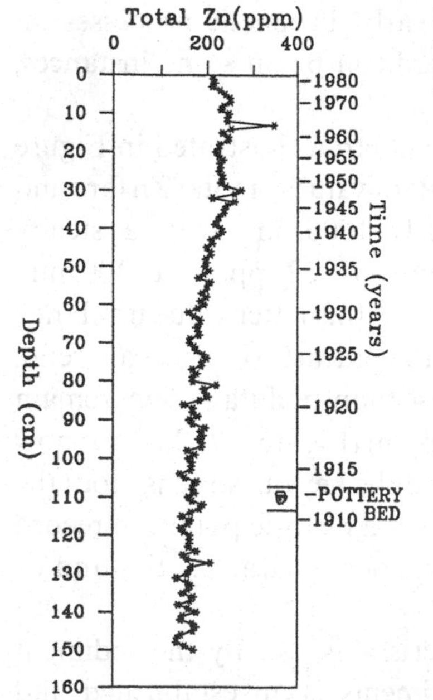

Figure 2.6 Variation in the saltmarsh sediment concentration of total zinc with depth in a dated (^{210}Pb and archaeological evidence) profile from Tites Point, Severn Estuary, UK.

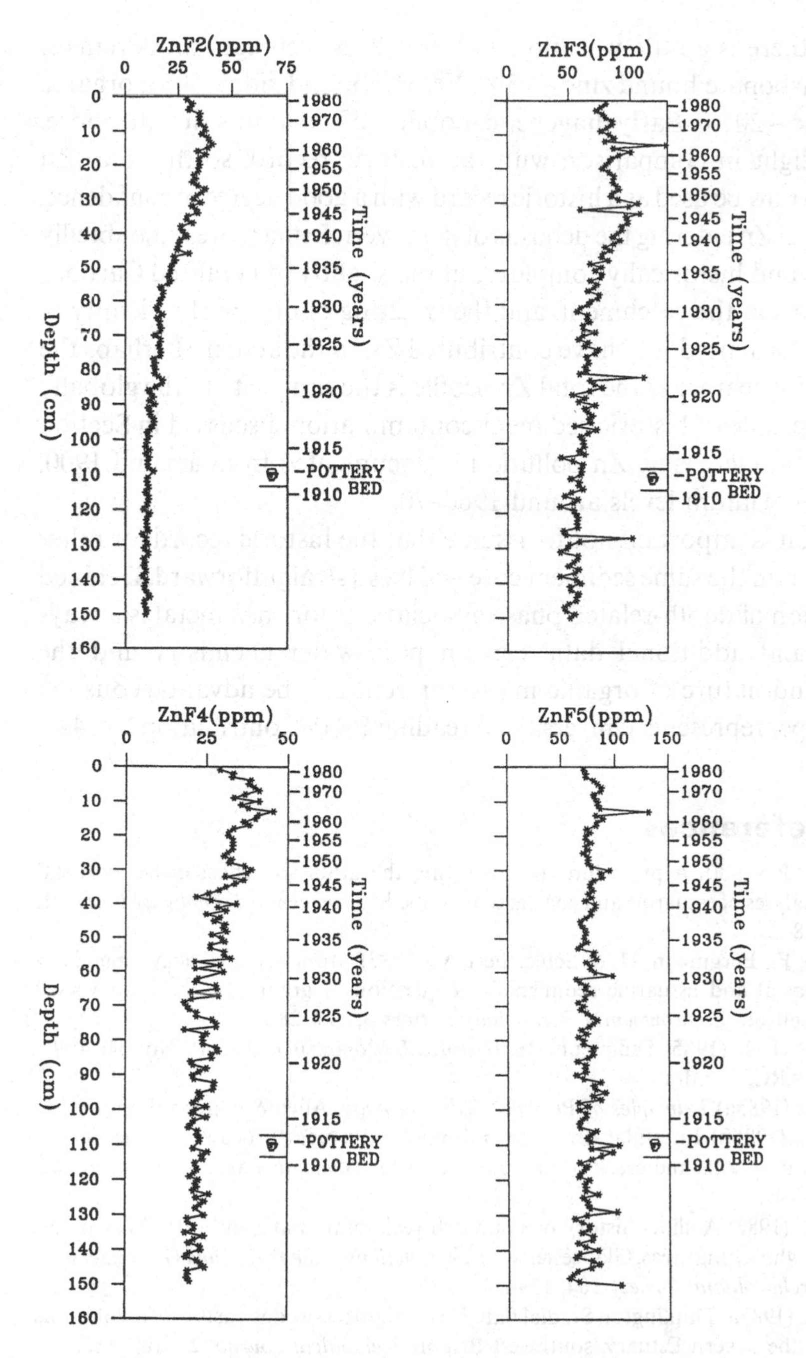

Figure 2.7 Variation in the saltmarsh sediment concentration of zinc with depth in four fractions, defined according to the sequential extraction scheme of Tessier, Campbell & Bisson (1979) as: F2, carbonate; F3, Fe and Mn oxide and hydroxide; F4, organic; F5, residual. The location of the profile is Tites Point, Severn Estuary, UK.

although there is a little variation with depth, percentage values remain similar (carbonate bound zinc ~ 15%, Fe/Mn bound zinc ~ 60%, organic bound zinc ~ 20%). Early diagenetic mobility of Zn at this site, therefore, appears slight in comparison with the historic record, so the total Zn profile can now be used as a historic record with a good degree of confidence.

Sources of Zn entering the deposits of the Severn Estuary are undoubtedly numerous and historically complex, but the slightly mineralised Carboniferous coals in the catchment, and the smelting of ores in the vicinity of the estuary will no doubt have contributed Zn to the system. Perhaps the most striking feature of the total Zn profile is the similarity to the globally averaged profile of historic sediment contamination discussed in Section 2.1.2; that is, increasing Zn pollution is documented from around 1900, reaching maximum levels around 1960–70.

Finally, it is important not to assume that the historic record for other trace metals in the same sediment core will be so straightforward. Detailed examination of depth-related phase associations for *each* metal is always essential and additional data, e.g. on pore-water chemistry and the amount and nature of organic matter present can be advantageous.

This paper represents University of Reading PRIS Contribution No. 443.

References

Ackermann, F. (1980) A procedure for correcting the grain size effect in heavy metal analyses of estuarine and coastal sediments. *Environmental Technology Letters* **1**, 518–27.

Ackermann, F., Bergmann, H. & Schleichert, V. (1983) Monitoring of heavy metals in coastal and estuarine sediments – a question of grain size: < 20 μm versus < 60 μm. *Environmental Technology Letters* **4**, 317–28.

Alderton, D.H.M. (1985) Sediments. In *Historical Monitoring*, Report No. 31, 1–95, MARC, London.

Allen, J.R.L. (1985a) *Principles of Physical Sedimentology*, Allen & Unwin, London.

Allen, J.R.L. (1985b) Intertidal drainage and mass-movement processes in the Severn Estuary: rills and creeks (pills). *Journal of the Geological Society*, London **142**, 849–61.

Allen, J.R.L. (1986) A short history of saltmarsh reclamation at Slimbridge Warth and neighbouring areas, Gloucestershire. *Transactions of the Bristol and Gloucestershire Archaeologial Society* **104**, 139–55.

Allen, J.R.L. (1987a) Dimlington Stadial (late Devensian) ice-wedge casts and involutions in the Severn Estuary, southwest Britain. *Geological Journal* **22**, 109–18.

Allen J.R.L. (1987b) Late Flandrian shoreline oscillations in the Severn Estuary: the Rumney Formation at its typesite (Cardiff area). *Philosophical Transactions of the Royal Society*, B **315**, 157–84.

Allen, J.R.L. (1987c) Towards a quantitative chemostratigraphic model for the sediments of late Flandrian age in the Severn Estuary, U.K. *Sedimentary Geology* **53**, 73–100.

Allen, J.R.L. (1987d) Reworking of muddy intertidal sediments in the Severn Estuary, Southwestern U.K. – a preliminary study. *Sedimentary Geology* **50**, 1–23.

Allen, J.R.L. (1988) Modern-period muddy sediments in the Severn Estuary (southwestern U.K.): a pollutant-based model for dating and correlation. *Sedimentary Geology* **58**, 1–21.

Allen, J.R.L. (1990a) The Severn Estuary in southwest Britain: its retreat under marine transgression, and fine-sediment regime. *Sedimentary Geology* **96**, 13–28.

Allen, J.R.L. (1990b) Reclamation and sea defence in Rumney Parish (Monmouthshire). *Archaeologia Cambrensis* **97**, 135–40.

Allen, J.R.L. (1990c) Late Flandrian shoreline oscillations in the Severn Estuary: change and reclamation at Arlingham, Gloucestershire. *Philosophical Transactions of the Royal Society*, A **330**, 315–34.

Allen, J.R.L. (1990d) Saltmarsh growth and stratification: a numerical model with special reference to the Severn Estuary, southwest Britain. *Marine Geology* **95**, 77–96.

Allen, J.R.L. (1990e) Constraints on measurement of sea-level movements from saltmarsh accretion rates. *Journal of the Geological Society*, London **147**, 5–7.

Allen, J.R.L. (1990f) The formation of coastal peat marshes under an upward tendency of relative sea-level. *Journal of the Geological Society*, London **147**, 743–5.

Allen, J.R.L. (1991a) Saltmarsh accretion and sea-level movement in the inner Severn Estuary: the archaeological and historical contribution. *Journal of the Geological Society*, London **148**, 485–94.

Allen, J.R.L. (1991b) Fine sediment and its sources, Severn Estuary and inner Bristol Channel, southwest Britain. *Sedimentary Geology* **75**, 57–65.

Allen, J.R.L. (1992a) Large scale textural patterns and sediment processes on tidal saltmarshes in the Severn Estuary, southwest Britain. *Sedimentary Geology* **81**, 299–318.

Allen, J.R.L. (1992b) Tidally influenced marshes in the Severn Estuary, southwest Britain. In J.R.L. Allen & K. Pye (eds.), *Saltmarshes: Morphodynamics, Conservation and Engineering Significance*, Cambridge University Press, Cambridge, 123–47.

Allen, J.R.L. & French, P.W. (1989) An apparatus for sequentially sampling unconsolidated cohesive sediments exposed on saltmarsh and river cliffs. *Sedimentary Geology* **61**, 151–4.

Allen, J.R.L. & Rae, J.E. (1986) Time sequence of metal pollution, Severn Estuary, southwestern U.K. *Marine Pollution Bulletin* **17**, 427–31.

Allen, J.R.L. & Rae, J.E. (1987) Late Flandrian shoreline oscillations in the Severn Estuary: a geomorphological and stratigraphical reconnaissance. *Philosophical Transactions of the Royal Society* B **315**, 185–230.

Allen, J.R.L. & Rae, J.E. (1988) Vertical saltmarsh accretion since the Roman period in the Severn Estuary, southwestern Britain. *Marine Geology* **83**, 225–35.

Allen, J.R.L., Rae, J.E. & Zanin, P.E. (1990) Metal speciation (Cu, Zn, Pb) and organic matter in an oxic salt marsh. Severn Estuary, southwest Britain. *Marine Pollution Bulletin* **21**, 574–80.

Allen, J.R.L., Rae, J.E., Longworth, G., Hasler, S.E. & Ivanovich, M. (1993) A comparison of the ^{210}Pb dating technique with three other independent dating methods in an oxic estuarine saltmarsh sequence. *Estuaries* **16**, 670–7.

Allison, J.A. & Perdue, E.M. (1994) Modelling metal-humic interactions with MINTEQA2. *Proceedings of the 6th International Meeting of the International Humic Substances Society, Bari, Italy.*

Araujo, M.F., Bernard, P.C. & Van Erieken, R.E. (1988) Heavy metal contamination in sediments from the Belgian Coast and Scheldt Estuary. *Marine Pollution Bulletin* **19**, 269–73.

Banat, K., Forstner, U. & Muller, G. (1972) Schwermetalle in den Sedimentendes Rheinso Unschau **72**, 192–3.

Bricker, S.B. (1990) The history of metals pollution in Narragansett Bay as recorded by saltmarsh sediments. PhD Dissertation. University of Rhode Island, Kingston, Rhode Island, USA.

Bryan, G.W. (1985) Bioavailability and effects of heavy metals in marine deposits. In B.H. Ketchum, J.M. Capuzzo, W.V. Burt, I.W. Duedall, P.K. Park & D.R. Kestner (eds.), *Wastes in the Ocean Vol 6 Nearshore Waste Disposal*, John Wiley & Sons, New York, 42–79.

Bryan, G.W. & Langston, W.J. (1992) Bioavailability, accumulation and effects of heavy metals in sediments with special reference to United Kingdom estuaries: a review. *Environmental Pollution* **76**, 89–131.

Buckley, D.E. & Winters, G.V. (1992) Geochemical characteristics of contaminated surficial sediments in Halifax Harbour: impact of waste discharge. *Canadian Journal of Earth Sciences* **29**, 2617–39.

Buckley, D.E., Smith, J.N. & Winters, G.V. (1995) Accumulation of contaminant metals in marine sediments of Halifax Harbour, Nova Scotia: environmental factors and historical trends. *Applied Geochemistry* **10**, 175–95.

Chester, R. & Hughes, M.J. (1967) A chemical technique for the separation of ferro-manganese minerals, carbonate minerals and adsorbed trace elements from pelagic sediments. *Chemical Geology* **2**, 249–62.

Cortesäo, C. & Vale, C. (1995) Metals in sediments of the Sado Estuary, Portugal. *Marine Pollution Bulletin* **30**, 34–7.

Crickmore, M.J. (1982) Data collection – tides, tidal currents, suspended sediment. In Institution of Civil Engineers (eds.), *Severn Barrage*, Thomas Telford, London, 19–26.

de Groot, A.J., Salomons, W. & Allersma, E. (1976) Processes affecting heavy metals in estuarine sediments. In J.D. Burton and P.S. Liss (eds.), *Estuarine Chemistry*, Academic Press, London, 131–57.

Din, Z. (1992) Use of aluminium to normalize heavy metals data from the estuarine and coastal sediments of the straits of Melaka. *Marine Pollution Bulletin* **24**, 484–91.

Farmer, J.G. (1991) The pertubation of historical pollution records in aquatic sediments. *Environmental Geochemistry and Health* **13**, 76–83.

Fengler, G., Grossman, D., Kersten, M. & Liebezeit, G. (1994) Trace metals in humic acids from recent Skagerrah sediments. *Marine Pollution Bulletin* **28**, 143–7.

French, P.W. (1993a) Areal distribution of selected pollutants in contemporary intertidal sediments of the Severn Estuary and Bristol Channel, U.K. *Marine Pollution Bulletin* **26**, 692–7.

French, P.W. (1993b) Seasonal and inter-annual variation of selected pollutants in modern intertidal sediments, Aust Cliff, Severn Estuary. *Estuarine, Coastal and Shelf Science* **37**, 213–19.

Gaudette, H.E., Flight, W.R., Toner, L. & Folger, D.W. (1974) An inexpensive titration method for the determination of organic carbon in recent sediments. *Journal of Sedimentary Petrology* **44**, 249–53.

Goede, A.A. (1985) Mercury, selenium, arsenic and zinc in waters from the Dutch Wadden Sea. *Environmental Pollution* **37A**, 287–309.

Goldberg, E.D. (1963) Geochronology with ^{210}Pb. In *Radioactive Dating*, International Atomic Energy Agency, Vienna.

Goldberg, E.D., Griffin, J.J., Hodge, V., Koide, M. & Windom, H. (1979) Pollution history of the Savannah River Estuary. *Environmental Science and Technology* **3**, 588–94.

Hem, J.D. (1972) Chemistry and occurrence of cadmium and zinc in surface water and groundwater. *Water Resources Research* **8**, 661–79.

Kheboian, C. & Bauer, C.F. (1987) Accuracy of selective extraction procedures for metal speciation in model aquatic sediments. *Analytic Chemistry* **59**, 1417–23.

Langston, W.J. (1980) Arsenic in UK estuarine sediments and its availability to benthic organisms. *Journal of the Marine Biological Association* **60**, 869–81.

Langston, W.J. (1982) The distribution of mercury in British estuarine sediments and its availability to deposit feeding bivalves. *Journal of the Marine Biological Association* **62**, 667–84.

Langston, W.J. (1986) Metals in sediments and benthic organisms in the Mersey estuary. *Estuarine Coastal & Shelf Science* **23**, 239–61.

Law, A.T. & Singh, A. (1988) Heavy metals in the Kelang estuary, Malaysia. *Malayan N. Journal* **41**, 505–13.

Loring, D.H. (1978) Geochemistry of zinc, copper and lead in the sediments of the estuary and Gulf of St. Lawrence. *Canadian Journal of Earth Science* **15**, 757–72.

Loring, D.H. (1990) Lithium – a new approach for the granulometric normalization of trace metal data. *Marine Chemistry* **29**, 155–68.

Luoma, S.N. (1989) Can we determine the biological availability of sediment-bound trace metals? *Hydrobiologia*, **176/177**, 379–96.

Luoma, S.N. & Bryan, G.W. (1978) Factors controlling the availability of sediment-bound lead to the estuarine bivalve *Scrobicularia plana*. *Journal of the Marine Biological Association* **58**, 793–802.

Luoma, S.N. & Bryan, G.W. (1982) A statistical study of environmental factors controlling concentrations of heavy metals in the burrowing bivalve *Scrobicularia plana* and the polychaete *Nereis diversicolor*. *Estuarine Coastal & Shelf Science* **15**, 95–108.

Luoma, S.N. & Phillips, D.J.H. (1988) Distribution, variability and impact of trace elements in San Francisco Bay. *Marine Pollution Bulletin* **19**, 413–25.

Mackey, A.P., Hodgkinson, M. & Nardella, R. (1992) Nutrient levels and heavy metals in mangrove sediments from the Brisbane River, Australia. *Marine Pollution Bulletin* **24**, 418–20.

McCaffrey, R.J. (1977) A record of the accumulation of sediment and trace metals in a Connecticut, U.S.A. saltmarsh. PhD Thesis, Yale University, New Haven, Connecticut, USA.

McCaffrey, R.J. & Thomson, J. (1980) A record of the accumulation of sediment and trace metals in a Connecticut saltmarsh. *Advances in Geophysics* **22**, 165–236.

Martin, J.M., Nirel, P. & Thomas, A.J. (1987) Sequential extraction techniques: promises and problems. *Marine Chemistry* **22**, 313–41.

Mattigod, S.V. & Sposito, G. (1979) Chemical modelling of trace metal equilibria in contaminated soil solutions using the computer programme GEOCHEM. In E.A. Jenne (ed.), *Chemical Modelling in Aqueous Systems*, American Chemical Society Symposium, Washington DC, 837–56.

Morris, A.W. (1984) The chemistry of the Severn Estuary and the Bristol Channel. *Marine Pollution Bulletin* **15**, 57–61.

Morrisey, D.J., Underwood, A.J., Stark, J.S. & Hewitt, I. (1994) Temporal variation in concentrations of heavy metals in marine sediments. *Estuarine Coastal and Shelf Science* **38**, 271–82.

Nirel, P.M. & Morel, F.M. (1990) Pitfalls of sequential extractions. *Water Research* **24**, 1055–6.

Oliver, B.G. (1973) Heavy metal levels of Ottowa and Rideau river sediments. *Environmental Science and Technology* **7**, 135–7.

O'Reilly Wiese, S.B., Bubb, J.M. & Lester, J.N. (1995) The significance of sediment metal concentrations in two eroding Essex salt marshes. *Marine Pollution Bulletin* **30**, 190–9.

Owens, M. (1984) Severn Estuary – an appraisal of water quality. *Marine Pollution Bulletin* **15**, 41–7.

Pickering, W.F. (1981) Selective chemical extraction of soil components and bound metal species. CRC Critical Reviews. *Analytical Chemistry* **12**, 233–66.

Radford, P.J. (1982) The effect of a barrage on water quality. In Institution of Civil Engineers (eds.), *Severn Barrage*, Thomas Telford, London, 203–8.

Rae, J.E. (1980) The geochemistry of mercury in estuarine mixing and sedimentation. PhD Thesis, University of Lancaster, Lancaster, UK.

Rae, J.E. & Allen, J.R.L. (1993) The significance of organic matter degradation in the interpretation of historical pollution trends in depth profiles of estuarine sediment. *Estuaries* **16**, 678–82.

Rowlatt, S.M. & Lovell, D.R. (1994) Lead, zinc and chromium in sediments around England and Wales. *Marine Pollution Bulletin* **28**, 324–9.

Salomons, W. & Mook, W.G. (1977) Trace metal concentrations in estuarine sediments: mobilization, mixing or precipitation. *Netherlands Journal of Sea Research*, **11**, 119–29.

Saxby, J.D. (1969) Metal-organic chemistry of the geochemical cycle. *Reviews of Pure and Applied Chemistry* **19**, 131–49.

Sharma, P.T., Church, M., Murray, S. & Biggs, R.B. (1987) Geochronology and trace metal records in a Delaware saltmarsh sediment. *Eos* **68**, 331.

Shuttler, R.M. (1982) The wave climate in the Severn Estuary. In Institution of Civil Engineers (eds.), *Severn Barrage*, Thomas Telford, London, 27–34.

Susetyo, W., Carreira, L.A., Azarraga, L.V. & Grimm, D.M. (1991) Fluorescence techniques for metal–humic interactions. *Fres. Zeit Analytical Chemistry* **339**, 624–35.

Tessier, A. & Campbell, P.G.C. (1987) Partitioning of trace metals in sediments: relationships with bioavailability. *Hydrobiologia* **149**, 43–52.

Tessier, A., Campbell, P.G.C. & Bisson, M. (1979) Sequential extraction procedure for the speciation of particulate trace metals. *Analytical Chemistry* **51**, 844–51.

Tessier, A., Campbell, P.G.C., Auclair, J.C. & Bisson, M. (1984) Relationships between the partitioning of trace metals in sediments and their accumulation in the tissues of the freshwater mollusc *Elliptio complanata* in a mining area. *Canadian Journal of Fisheries and Aquatic Science* **41**, 1463–72.

Thomas, R.L. & Jaquet, J.M. (1976) Mercury in surficial sediments of Lake Erie. *Journal of the Fisheries Research Board of Canada* **33**, 404–12.

Thomson-Becker, E.A. & Luoma, S.N. (1985) Temporal fluctuations in grain size, organic materials and iron concentrations in intertidal surface sediment of San Francisco Bay. *Hydrobiologia* **129**, 91–107.

Tipping, E. (1993a) Modelling the competition between alkaline earth cations and trace metal species for binding by humic substances. *Environmental Science and Technology* **27**, 520–9.

Tipping, E. (1993b) Modelling ion-binding by humic acids. *Colloid Surfaces* **73**, 117–31.

Tipping, E. (1994) WHAM – A chemical equilibrium model and computer code for waters, sediments and soils incorporating a discrete site/electrostatic model of ion-binding by humic substances. *Computers and Geosciences* **20**, 973–1023.

Tipping, E. & Hurley, M.A. (1992) A unifying model of cation binding by humic substances. *Geochimica et Cosmochimica Acta* **56**, 3627–41.

Tipping, E. & Woof, C. (1990) Humic substances in acid organic soils: modelling their release to the soil solution in terms of humic charge. *Journal of Soil Science* **41**, 573–83.

Tipping, E. & Woof, C. (1991) The distribution of humic substances between the solid and aqueous phases of acid organic soils; a description based on humic heterogeneity and charge-dependent sorption equilibria. *Journal of Soil Science* **42**, 437–48.

Towner, J.V. (1984) Studies of chemical extraction techniques used for elucidating the partitioning of trace metals in sediments. PhD Thesis, University of Liverpool, Liverpool, UK.

Turner, A. & Millward, G.E. (1993) Application of the K_D concept to the study of trace metal removal and adsorption during estuarine mixing. *Estuarine, Coastal and Shelf Science* **36**, 1–13.

Uncles, R.J. (1984) Hydrodynamics of the Bristol Channel. *Marine Pollution Bulletin* **15**, 47–53.

Ure, A.M., Quevauviller, Ph., Muntau, H. & Griepink, B. (1993) Speciation of heavy metals in soils and sediments. An account of the improvement in harmonisation of extraction techniques under the auspices of the BCR of the Commission of the European Union. *International Journal of Environmental Analytical Chemistry* **51**, 135–51.

Valette-Silver, N.J. (1992) Historical reconstruction of contamination using sediment cores: A review. *NOAA Technical Memorandum NOS.ORCA 65*, Rockville, Maryland.

Valette-Silver, N.J. (1993) The use of sediment cores to reconstruct historical trends in contamination of estuarine and coastal sediments. *Estuaries* **16**, 577–88.

Whalley, C. & Grant, A. (1994) Assessment of the phase selectivity of the European Community Bureau of Reference (BCR) sequential extraction procedure for metals in sediment. *Analytical Chimica Acta* **291**, 287–95.

Windom, H.L., Schropp, S.J., Calder, F.D., Ryan, J.D., Smith, R.G. Jr, Burney, L.C., Lewis, F.G. & Rawlinson, C.H. (1989) Natural trace metal concentrations in estuarine and coastal sediments of the southeastern United States. *Environmental Science and Technology* **23**, 314–20.

Zwolsman, J.J.G., Berger, G.W. & Van Eck, G.T.M. (1993) Sediment accumulation rates, historical input, postdepositioned mobility and retention of major elements and trace metals in salt marsh sediments of the Scheldt estuary, SW Netherlands. *Marine Chemistry* **44**, 73–94.

3

○ ○ ○ ○ ○ ○ ○ ○ ○ ○ ○ ○ ○ ○ ○ ○ ○ ○ ○ ○

Modelling adsorption and desorption processes in estuaries

A. Turner and A.O. Tyler

3.1 Introduction

Many inorganic and organic pollutants are of great concern to water quality managers owing to their persistence, toxicity and liability to bioaccumulate (Tanabe, 1988; Robards & Worsfold, 1991; Bryan & Langston, 1992; McCain *et al.*, 1992; Förstner, 1993). The major temporary or ultimate sink for such pollutants in estuaries is the sedimentary reservoir, including intertidal deposits (Hanson, Evans & Colby, 1993; van Zoest & van Eck, 1993; Kennicutt *et al.*, 1994), and definition of the biogeochemical mechanisms by which they absorb onto, desorb from and repartition amongst natural, heterogeneous particle populations is essential in order to assess their environmental fate. In estuaries, prediction of the distribution of pollutants is further compounded by intense temporal and spatial gradients of the reaction controlling variables such as salinity, dissolved oxygen concentration and particle composition, occurring both within the sediment and in the water column during particle suspension.

This chapter focuses on an empirical technique to study the sorptive behaviour of trace metals and trace organic compounds in estuaries. Thus, the partitioning of constituents between particles and solution is determined experimentally, without identifying the inherent reaction mechanisms or reactant speciation, under controlled laboratory conditions using natural samples spiked with radiotracer analogues of the constituent of interest (Turner *et al.*, 1993). Adsorption and desorption processes may be modelled as a function of particle concentration, and the controlled variables of salinity and dissolved oxygen concentration, by incorporating empirically derived results into simple mass balance equations. This approach is exemplified herein using site-specific results from two contrasting estuarine environments, namely the Clyde and Humber, and the calculated results are discussed in the context of their agreement with field measurements of trace constituents in these estuaries. Although the arguments and discussion are presented explicitly for sorption processes occurring in the water column they may be extrapolated to the intertidal environment where chemical reactivity is also promoted by changes in dissolved oxygen concentration (Kerner & Wallmann, 1992) and salinity (Carroll *et al.*, 1993).

3.1.1 The Clyde and Humber Estuaries

The physical characteristics of the Clyde and Humber Estuaries are listed in Table 3.1, and typical axial distributions of salinity, concentration of suspended particulate matter (SPM) and dissolved oxygen concentration are shown in Figure 3.1. The Clyde is a partially stratified, mesotidal estuary. Irregularities in its axial salinity distribution arise from a multiplicity of fresh water sources along the tidal estuary. The SPM concentrations are low and relatively invariant and an occasional, weak turbidity maximum results from trapping of fine sediment by convergent subsurface flows (Curran, 1986). The high biochemical oxygen demand of sewage discharges combined with the poor exchange between subsurface and surface waters results in an extensive area of oxygen depletion in the upper estuary (Curran, 1986). In contrast, the Humber is a well-mixed, macrotidal estuary, and a turbidity maximum generated by tidal resuspension of bed sediment is a regular feature of the low salinity zone (Turner, Millward & Morris, 1991).

3.2 Determination of distribution coefficients

The distribution coefficient, or partition coefficient, K_D (v/w), parameterises the ratio of adsorbed particulate concentration, P (w/w), to dissolved concentration, C (w/v), of a constituent, and may be determined empirically

Table 3.1 Physical characteristics of the Clyde and Humber Estuaries

	Clyde[a]	Humber[b]
Axial length (km)	40	62
Mean river flow ($m^3\ s^{-1}$)	44 (+63[c])	246
Tidal range at mouth, neaps/spring (m)	1.9/3.0	3.5/6.2
Residence time (days)	10–20	~40
Riverine sediment discharge ($t\ a^{-1}$)	83 000	170 000

[a] Mackay & Leatherland (1976); Curran (1986)
[b] Turner *et al.* (1991)
[c] contribution from tributaries to the tidal estuary

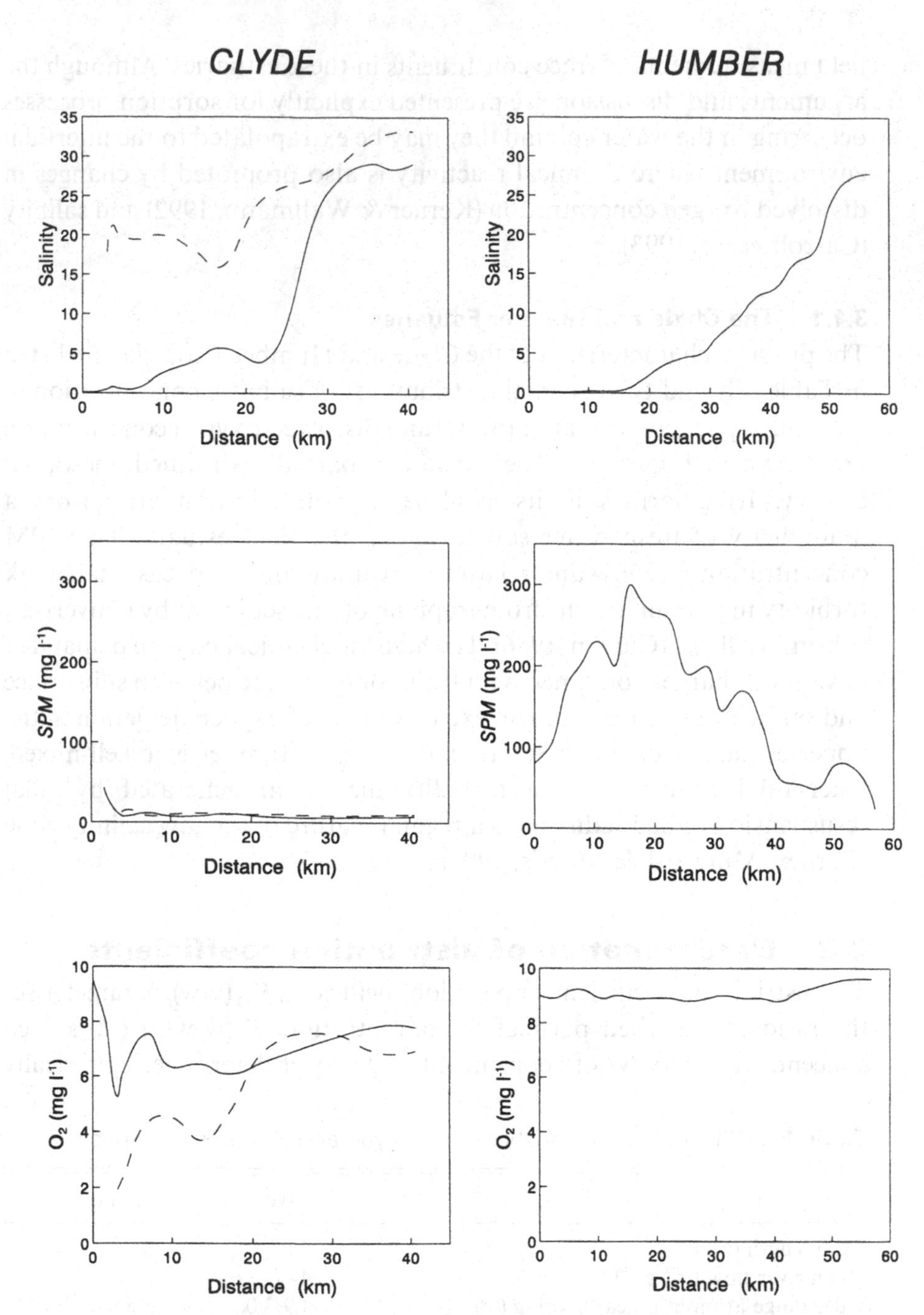

Figure 3.1 Distributions of salinity, suspended particulate matter (SPM) and dissolved oxygen determined during axial samplings of the Clyde (8/92; ebb tide, river flow including tributary inputs = 152 m^3 s^{-1}) and Humber (1/88; flood tide, river flow = 234 m^3 s^{-1}). Solid lines denote near surface measurements and broken lines denote near bottom measurements. Distance is measured from the weir at Glasgow on the Clyde and from Trent Falls on the Humber.

by doping natural samples with radiotracer analogues and analysing the adsorbed and dissolved reactant concentrations by a suitable radiochemical method.

Water samples used for the experimental studies of trace metals and chlorinated organic compounds were collected from the shore near to the river and marine end-members of the Humber and Clyde Estuaries and stored cool and in the dark before being used. Trace metal studies using Humber samples were undertaken onboard Royal Research Ship *Challenger* within 48 hours of collection and are described in detail elsewhere (Turner *et al.*, 1993); the remaining experiments were undertaken in the laboratory in Plymouth within one week of sample collection. The simulation of estuarine mixing was achieved by mixing aliquots of unfiltered end-member samples in different proportions. For the trace metal studies mixed samples of 50 or 100 ml were contained within acid-cleaned 150 ml polyethylene vessels and spiked with 50 or 100 μl of a cocktail consisting of sub-nanomolar concentrations of ^{109}Cd and ^{137}Cs (Amersham International) in dilute HCl. This resulted in a small reduction in pH in the Humber experiments which was neutralised by the addition of a spike of ammonia solution diluted accordingly. The samples were equilibrated for 120 hours and filtered through pre-weighed 0.45 μm pore size Millipore cellulose acetate filters mounted in a Millipore filtration unit. The filtrates and filters were stored in unused 150 ml polyethylene vessels and 50 mm diameter petri dishes, respectively, and their activities were determined using a high purity Ge detector connected to a Canberra Series 80 Multichannel Analyser. Disintegrations were counted for 1000 seconds and this gave statistical counting errors ($\pm 0.5\,\sigma$) of $\lesssim 5\%$. The distribution coefficient, K_D (ml g^{-1}), is then given as follows:

$$K_D = \frac{P}{C} = \frac{A_p V}{A_c mf} \tag{3.1}$$

where A_p and A_c are the activities on the particles and in solution, respectively, m is the mass of particles on the filter (g), V is the volume of filtrate (ml), and f is a geometry factor which corrects for the differential sensitivity of the detector towards filter and filtrate samples. Replicate experiments using estuarine samples have indicated that K_Ds determined by this technique are generally reproducible within 10% (Turner & Millward, 1994).

The effect of oxygen depletion on the partitioning was investigated by bubbling nitrogen through a Clyde river end-member sample for 12 hours prior to spiking and thereafter throughout the incubation. This achieved a dissolved oxygen concentration of 1.5 mg l^{-1} compared with 8.0–8.5 mg l^{-1} under ambient, aerated laboratory conditions, and thus approximates the observed range in concentrations of the upper reaches of the estuary (Figure 3.1).

Two ^{14}C-labelled chlorinated organic compounds were selected for investigation: 2,3,7,8-TCDD (Chemsyn), the most toxic of the dioxin congeners; and 22′55′-TCB, IUPAC 52 (Sigma Chemicals), an abundant PCB congener. A 25 μl spike of 0.02 μCi compound in hexane was placed on the side of a hexane-washed 50 ml glass centrifuge tube using a glass microsyringe; the hexane was then evaporated under a laminar flow hood and a sample of 20 ml was added (Zhou, Rowland & Mantoura, 1995; Zhou & Rowland, 1997). After shaking the stoppered tube for 12 hours, the contents were centrifuged at 3000 rpm for 30 minutes to separate the particulate and dissolved phases. A 2 ml aliquot of supernatant was pipetted into a glass vial containing 10 ml Ultima Gold Liquid Scintillation Cocktail, taking care not to disturb particles at the bottom of the centrifuge tube. The particles were then discarded along with the remaining water and the residual compound adsorbed to the centrifuge walls was extracted in 4 ml hexane for 12 hours on a shaker. A 2 ml aliquot of the extract was then pipetted into a glass vial containing 10 ml scintillation cocktail. All sample vials were counted on a Philips 4700 scintillation counter and calibrated against matrix-matched standards quenched using CCl_4. The K_D (ml g^{-1}) was derived from mass balance as follows:

$$K_D = \frac{P}{C} = \frac{A_s - (A_c + A_w)}{A_c} \cdot \frac{V}{m} \tag{3.2}$$

where A_s, A_c and A_w are the respective activities of the original spike, the supernatant and that adsorbed onto the walls of the centrifuge tube, V is the volume of supernatant (ml) and m is the mass of particles (g) in the original suspension derived from subsample filtration through pre-weighed Whatman GF/F filters. Experiments were performed in triplicate or quadruplicate and the reproducibility of the resulting K_Ds was generally better than 10% for 22′55′-TCB, and up to about 40% for 2,3,7,8-TCDD due to its poor solubility and consequent low activity in solution.

3.3 Results and model calculations

3.3.1 Fraction of constituent in solution

The K_Ds for the trace metals and chlorinated organic compounds are shown against salinity in Figure 3.2. End-member K_Ds have been combined with representative end-member SPM concentrations (SPM) (Figure 3.1) in order to calculate the fraction of constituent in solution, ζ, as follows:

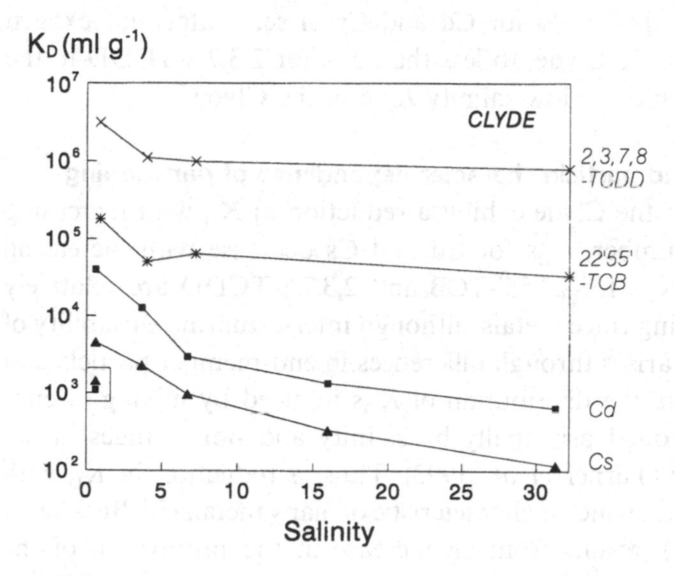

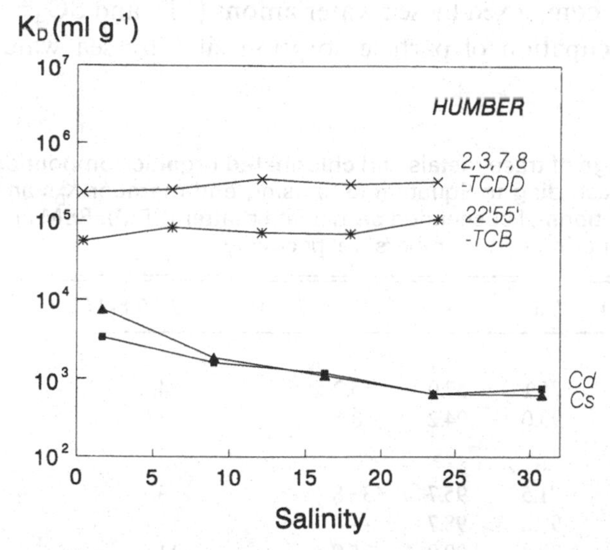

Figure 3.2 Distribution coefficients, K_Ds, against salinity for the Clyde and Humber mixed samples. Boxed data points denote K_Ds determined using the sample incubated under de-oxygenated conditions.

$$\zeta = \frac{C}{C + P \cdot SPM} \quad (3.3)$$
$$= \frac{1}{1 + K_D \cdot SPM}$$

The magnitude of this fraction (Table 3.2), which has important implications concerning the bioavailability, transport and fate of these constituents (Balls, 1989; Pankow & MacKenzie, 1991; Webster & Ridgway, 1994) varies from more than 90% for Cd and Cs in sea water and oxygen depleted waters of the Clyde, to less than 5% for 2,3,7,8-TCDD in the Humber Estuary and the low salinity zone of the Clyde.

3.3.2 Salinity and particle character dependence of partitioning

All constituents in the Clyde exhibit a reduction in K_D with increasing salinity; in the Humber, K_Ds for Cd and Cs decrease with increasing salinity whereas K_Ds for 22′55′-TCB and 2,3,7,8-TCDD are relatively invariant. Regarding trace metals, although inter-estuarine variability of end-member K_Ds arises through differences in end-member particle and water composition, the distribution of K_Ds induced by mixing of end-members is controlled principally by salinity and not changes in net particle character (Turner *et al.*, 1993). Thus, a reduction in K_D with increasing salinity, S, which is characteristic of many metals (Li, Burkhardt & Teraoka, 1984), results from an increase in the proportion of the sorbable species being complexed by sea water anions (Cl^- and SO_4^{2-}) and an increasing occupation of particle sorption sites by sea water

Table 3.2 The percentage of trace metals and chlorinated organic compounds in solution, calculated according to Equation (3.3) using end-member K_Ds and representative concentrations of suspended particulate matter (SPM). REM and MEM denote river and marine end-members, respectively

	SPM (mg l^{-1})	Cd	Cs	22′55′-TCB	2,3,7,8-TCDD
Humber					
REM	100	75.1	57.0	15.2	4.2
MEM	100	93.0	94.2	8.5	3.0
Clyde					
REM	10	71.5	95.7	35.8	3.1
REM*	10	98.8	98.7	—	—
MEM	10	99.4	99.9	75.0	11.2

*De-oxygenated river end-member sample

cations (Na^+, K^+, Ca^{2+}, Mg^{2+}), and may be defined by the following general relationship (Bale, 1987; Turner & Millward, 1994):

$$\ln K_D = b \cdot \ln(S + 1) + \ln K_D^0 \quad (3.4)$$

where K_D^0, the partitioning in fresh water, and b are constants. Values of these constants for Cd and Cs in the Clyde and Humber have been derived from regressions of $\ln K_D$ versus $\ln(S + 1)$ and are given in Table 3.3. The parameterisation of the sorptive behaviour of constituents in the form of such generic relationships is extremely useful for the refinement of chemical codings in contaminant transport models (Plymsolve, 1991; Ng *et al.*, 1996).

Regarding chlorinated organic compounds, the information presented herein and elsewhere (Tyler *et al.*, 1994) suggests that particle character rather than salinity exerts the key influence on their estuarine distributions. The chemical characteristics of typical Clyde and Humber end-member particles are given in Table 3.4. Thus, Humber river and marine end-member particles have similar chemical characteristics engendering a relatively uniform distribution of K_D, whereas in the Clyde, riverine particles have a higher content of Fe–Mn hydroxides and carbon for the sorption of compounds and a reduction in K_D is observed as the proportion of riverine particles in the admixture decreases. More specifically, recent studies suggest that adsorption of trace organic compounds onto natural particles is controlled dominantly by the apolar lipid content (Preston & Raymundo, 1993; Tyler *et al.*, 1994). Higher absolute K_Ds in the Clyde than in the Humber are, therefore, reflected by inter-estuarine differences in particulate lipid content (Clyde sediment = 6.5 ± 5.0 mg g^{-1} cf. Humber sediment 1.6 ± 0.7 mg g^{-1}; Tyler *et al.*, 1994).

Table 3.3 Regression analyses of $\ln K_D$ versus $\ln(S+1)$ (n = 5) for Cd and Cs (see Equation (3.4))

Estuary		Slope, b	Intercept, K_D^0 (ml g^{-1})	r^2 (%)	p
Clyde	Cd	– 1.42	80 800	96.5	0.002
	Cs	– 1.25	11 000	95.8	0.002
Humber	Cd	– 0.653	6 000	92.8	0.005
	Cs	– 1.05	21 000	99.0	< 0.001

3.3.3 Desorption modelling

The foregoing discussion suggests that the estuarine distributions of trace organic compounds are determined largely by natural particle mixing processes and the rates at which sorbed compounds redistribute amongst different particle types. In contrast, the partitioning of trace metals in estuaries is controlled principally by salinity (Equation (3.4)) and their distributions will be determined to some extent by desorption from seaward fluxing particles. Assuming no loss or gain of seaward fluxing particles, and neglecting possible additional sources of trace metals such as pore-water infusion, this effect may be quantified from mass balance of constituent as follows (Li *et al.*, 1984):

$$C^{S1} + P^{S1}{\cdot}SPM^{f} = C^{S2} + P^{S2}{\cdot}SPM^{f} \tag{3.5}$$

where SPM^{f} is the concentration of seaward fluxing particles traversing an axial or vertical salinity gradient, S1–S2. Assuming that the sorption reaction is fully reversible, the relative changes in adsorbed particulate and dissolved concentrations along this salinity gradient are given as follows:

$$\frac{P^{S2}}{P^{S1}} = \frac{SPM^{f} + 1/K_{D}^{S1}}{SPM^{f} + 1/K_{D}^{S2}} \tag{3.6}$$

and:

$$\frac{C^{S2}}{C^{S1}} = \frac{P^{S2}}{P^{S1}}{\cdot}\frac{K_{D}^{S1}}{K_{D}^{S2}} \tag{3.7}$$

Table 3.4 The composition of typical end-member particles from the Clyde and Humber. Fe and Mn were determined following extraction by hydroxylamine hydrochloride–acetic acid. Carbon and nitrogen were determined using an elemental analyser. Specific surface area (SSA) was determined using a BET nitrogen adsorption technique (nd = not determined). REM and MEM denote river and marine end-members, respectively.

	Fe ($mg\ g^{-1}$)	Mn ($mg\ g^{-1}$)	Carbon (%)	Carbon/ Nitrogen	SSA ($m^2\ g^{-1}$)
Clyde					
REM	8.40	1.02	9.9	9.2	7.1
MEM	3.52	0.29	9.3	10.3	nd
Humber[a]					
REM	10.1	1.32	4.9	15.8	31.4
MEM	11.2	1.05	5.2	13.9	20.6

[a] Turner *et al.* (1993)

Note that these equations are also valid for trace organic compounds if the seaward fluxing particle population is chemically modified *in situ* through, for example, precipitation or dissolution of Fe–Mn oxides and/or organic coatings.

The relative change in dissolved or adsorbed concentration of trace metals due to the flux of particles along a dissolved oxygen gradient, O2–O1, may be calculated likewise using K_Ds determined under different oxygen conditions. Thus:

$$\frac{P^{O2}}{P^{O1}} = \frac{SPM^f + 1/K_D^{O1}}{SPM^f + 1/K_D^{O2}} \tag{3.8}$$

and:

$$\frac{C^{O2}}{C^{O1}} = \frac{P^{O2}}{P^{O1}} \cdot \frac{K_D^{O1}}{K_D^{O2}} \tag{3.9}$$

Figure 3.3 shows dissolved and adsorbed particulate concentration ratios for Cd as a function of seaward fluxing particle concentration, SPM^f, calculated according to Equations (3.6) and (3.7) using end-member K_Ds for the Humber and Clyde, and according to Equations (3.8) and (3.9) using K_Ds determined under extreme dissolved oxygen concentrations in the Clyde. Although for a given concentration, SPM^f, the relative change in dissolved concentration in the Clyde effected by variations in both salinity and dissolved oxygen is greater than the relative change in dissolved concentration in the Humber effected by variations in salinity, the magnitude of the advecting particle population is considerably smaller in the Clyde. Thus, for a representative concentration of SPM^f in the Clyde of ~10 mg l^{-1} (Figure 3.1), $C^{S2}/C^{S1} = C^{O2}/C^{O1} \sim 1.4$, $P^{S2}/P^{S1} \sim 0.02$ and $P^{O2}/P^{O1} \sim 0.04$. Particle fluxes in dynamic macrotidal estuaries are spatially and temporally variable (Morris *et al.*, 1986), but a hypothetical particle population of concentration 250 mg l^{-1} traversing the full salinity range of the Humber Estuary yields the following concentration ratios: $C^{S2}/C^{S1} = 1.5$, and $P^{S2}/P^{S1} = 0.35$. According to these calculations the largest relative change in Cd concentration occurs in Clyde suspended particles and results in a 25- to 50-fold reduction of the adsorbed particulate fraction.

3.3.4 Adsorption modelling

Sorptive removal of dissolved constituents is engendered by the addition of sorption sites. In estuaries this is achieved through tidally induced resuspension of bed particles that are depleted of the adsorbed constituent relative to equilibrium. This effect may be quantified by the following equation derived from mass balance (Morris, 1986):

$$\frac{C}{C^0} = \frac{1 + K_D \cdot SPM^0}{1 + K_D \cdot SPM^0 + K_D \cdot SPM^a(1-\alpha)} \tag{3.10}$$

where C/C^0 is the ratio of dissolved constituent concentration in turbidised water to that in ambient water (i.e. fractional removal), SPM^a is the concentration of added or resuspended particles, SPM^0 is the concentration of ambient suspended particles, and α defines the depletion

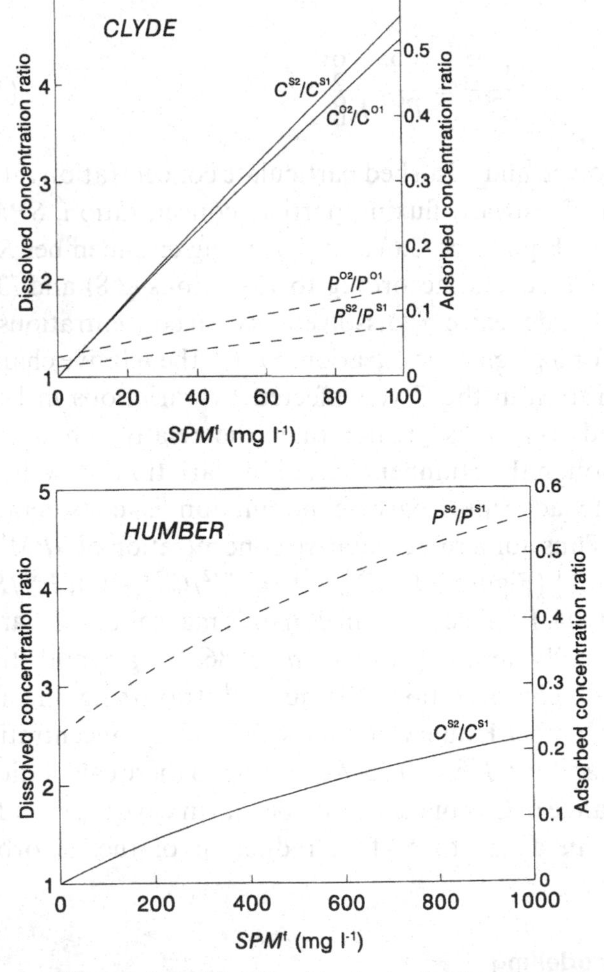

Figure 3.3 Dissolved and adsorbed particulate concentration ratios as a function of fluxing particle concentration (SPM^f) for Cd in the Clyde and Humber, calculated according to Equations (3.6)–(3.9).

of adsorbed constituent on the added particles relative to equilibrium in the turbidised water (i.e. P^a/P). Such constituent-depleted particles are derived from the marine reaches of an estuary where anthropogenic inputs are generally less significant and particle sorption sites are occupied by sea water cations, and are transported up-estuary by asymmetric tidal currents. The calculated sorptive removal of constituents is shown for the Humber Estuary in Figure 3.4 within an added particle concentration range of 0–1000 mg l^{-1}, using river end-member K_Ds and an ambient particle concentration of 5 mg l^{-1}; the value of α of 0.50 is based on empirical estimates given by Morris (1986). Under these conditions 50% removal of dissolved constituent is achieved at the following approximate added particle concentrations: Cd, 700 mg l^{-1}; Cs, 250 mg l^{-1}; 22′55′-TCB, 50 mg l^{-1}; 2,3,7,8-TCDD, 20 mg l^{-1}. Application of the sorption model to the Clyde Estuary is not appropriate for the following reasons. Firstly, although localised disturbance of the bed occurs during dredging operations, there is no evidence of extensive resuspension from axial and vertical SPM distributions (Figure 3.1). Moreover, bottom current speeds are lower than the critical erosion velocity for representative grain diameters (Curran, 1986). Secondly, there is little potential for up-estuary transport of marine particles due to slow subsurface currents and deep basins in the outer estuary which act as

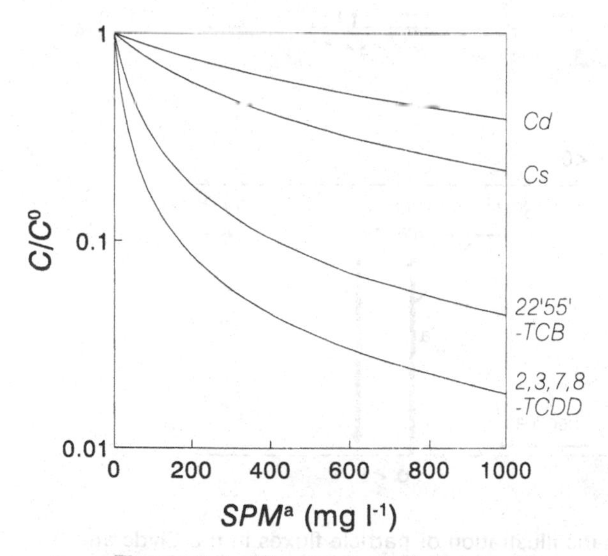

Figure 3.4 The removal of dissolved trace metals and chlorinated organic compounds as a function of added particle concentration (SPM^a) in the river end-member of the Humber Estuary, calculated according to Equation (3.10) for $SPM^0 = 5$ mg l^{-1} and $\alpha = 0.50$.

sediment traps (Mackay & Leatherland, 1976). It is conceivable, however, that removal of certain trace metals (including Cd) may be achieved via the precipitation of insoluble sulphide species (Zwolsman & van Eck, 1993). A continual source of sulphide (namely, the reducing bottom sediments) is not accounted for in the laboratory simulations and this serves to exemplify potential limitations with our attempts to replicate multiple, interactive environmental processes in laboratory enclosures.

3.4 Discussion

Although chemical parameters such as distribution coefficients are fundamental to understanding the geochemical reactivity of trace

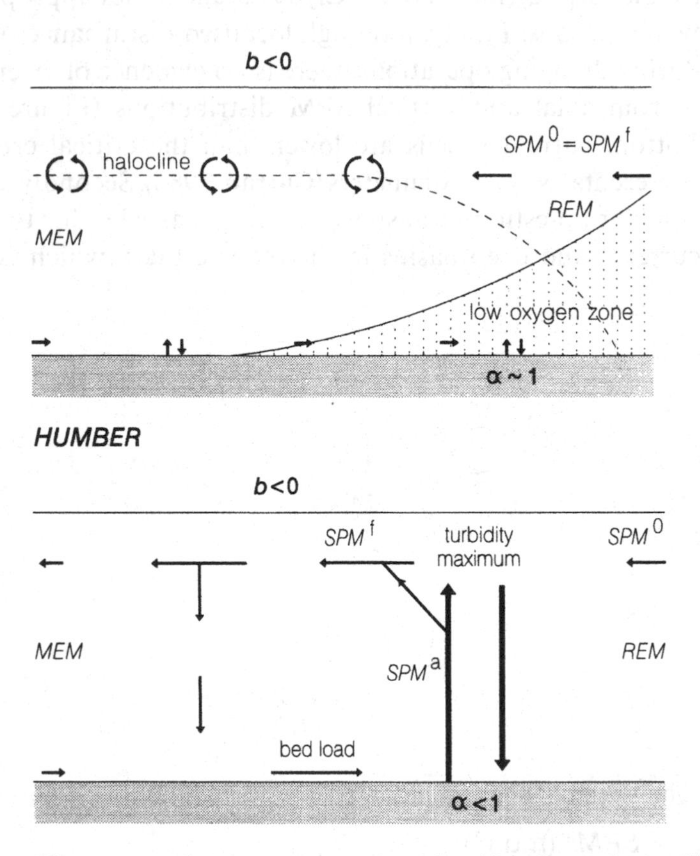

Figure 3.5 Schematic illustration of particle fluxes in the Clyde and Humber Estuaries. Arrow size denotes relative magnitude of the flux. The broken line represents the boundary of the saline intrusion and the river. REM and MEM denote river and marine end-members, respectively. All terms are defined in the text.

constituents in estuaries, the foregoing calculations have emphasised the important role of sediment dynamics on chemical distributions. In the Humber Estuary sorption reactions are promoted by substantial resuspending and advecting particle fluxes, as conceptualised in Figure 3.5, and the suspended particle populations are in a dynamic state of exchange with subtidal and intertidal material. In highly energetic, macrotidal estuaries a proportion of riverine suspended particles are transported into the turbidity maximum, where the magnitude of the seaward advecting population is augmented by particles resuspended by tidal currents (Morris *et al.*, 1986). Trace metals are desorbed from these particles downestuary in accordance with the magnitude of b, the rate of change of K_D with salinity, until they are deposited as the current speeds are reduced. An admixture of this sediment and marine derived material is then transported up-estuary as bedload by asymmetric tidal currents to the null zone where it is resuspended into the turbidity maximum engendering sorptive removal of dissolved constituent in accordance with the value of α, the extent of depletion (relative to equilibrium) of adsorbed constituent on the particles. Field investigations of various dissolved trace metals (diagnosed as constituent–salinity plots) in the Humber Estuary have demonstrated conservativeness (Balls, 1985; Kitts *et al.*, 1994) as well as a variety of reactive distributions (Gardiner, 1982; Edwards, Freestone & Urquhart, 1987; Coffey & Jickells, 1995). Differential behaviour amongst constituents reflects differences in chemical parameters such as end-member K_Ds, α, and the reversibility of sorption, whereas intra-constituent variability arises from temporal variability of component sediment fluxes.

Chemical reactivity in the Clyde is governed by the flux of particles through lateral and vertical gradients of salinity or dissolved oxygen and there is little exchange of particles between the water column and the bed (Figure 3.5). Because of the low and invariant concentrations of suspended particles, calculations predict that sorptive reactivity in the water column will result in a significant modification of the trace metal composition of suspended particles. Seasonal measurements of trace metals on suspended particles available to 1M NH_4OAc (an operational measure of an exchangeable particulate fraction) indicate a persistent reduction from the upper estuary to the lower estuary (Muller, Tranter & Balls, 1994); for Cd, an exchangeable particulate concentration ratio, X^{S2}/X^{S1}, is about 0.20. Although an order of magnitude greater than the adsorbed particulate ratio calculated (Equation (3.6)), reflecting analytical difficulties associated with the determination of an exchangeable fraction and the assumption in the calculations that the adsorbed metal is completely exchangeable (Equations (3.6)–(3.9)), these field measurements demonstrate a distinct seaward modification of the Cd content of the suspended particle population. Calculations also predict that accompanying changes in the dissolved phase are likely to be difficult to detect. Thus, conservative distributions of dissolved trace metals are generally observed (Mackay & Leatherland, 1976; Muller, Tranter & Balls, 1994), deviations from the theoretical dilution line being accounted for by multiple straight line segments resulting from a variety of tributary inputs of different compositions (Mackay & Leatherland, 1976), and direct anthropogenic inputs to the tidal estuary (Muller Tranter & Balls, 1994).

Although the arguments and calculations presented herein are qualitatively compatible with field observations from two contrasting estuaries, accurate quantitative predictions will rely on a better definition of chemical parameters such as α and sorption time constants, and a greater understanding of estuarine sediment dynamics including the interaction between suspended populations and subtidal/intertidal deposits. Nevertheless, the partitioning data and sorption models are particularly useful *per se* in providing a general understanding of the likely environmental transport and fate of constituents about which relatively little is known, such as PCBs and dioxins.

References

Bale, A.J. (1987) The characteristics, behaviour and heterogeneous reactivity of estuarine particles. Ph.D. Thesis, Polytechnic South West, Plymouth, UK, 216 pp.

Balls, P.W. (1985) Copper, lead and cadmium in coastal waters of the western North Sea. *Marine Chemistry* **15**, 363–78.

Balls, P.W. (1989) Trend monitoring of dissolved trace metals in coastal sea water – a waste of effort? *Marine Pollution Bulletin* **20**, 546–8.

Bryan, G.W. & Langston, W.J. (1992) Bioavailability, accumulation and effects of heavy metals in sediments with special reference to United Kingdom estuaries: a review. *Environmental Pollution* **76**, 89–131.

Carroll, J., Falkner, K.K., Brown, E.T. & Moore, W.S. (1993) The role of the Ganges–Brahmaputra mixing zone in supplying barium and ^{226}Ra to the Bay of Bengal. *Geochimica et Cosmochimica Acta* **57**, 2981–90.

Coffey, M.J. & Jickells, T.D. (1995) Ion chromatography–inductively coupled plasma–atomic emission spectrometry (IC-ICP-AES) as a method for determining trace metals in estuarine water. *Estuarine, Coastal and Shelf Science* **40**, 379–86.

Curran, J.C. (1986) Effluent disposal and the physical environment. *Proceedings of the Royal Society of Edinburgh* **90B**, 97–115.

Edwards, A., Freestone, R. & Urquhart, C. (1987) *The Water Quality of the Humber Estuary, 1986*. A Report of the Humber Estuary Committee, 27 pp.

Förstner, U. (1993) Metal speciation – general concepts and applications. *International Journal of Environmental and Analytical Chemistry* **51**, 5–23.

Gardiner, J. (1982) Nutrients and persistent contaminants. In A.L.H. Gameson (ed.) *The Quality of the Humber Estuary.* Yorkshire Water Authority, Leeds, 27–33.

Hanson, P.J., Evans, D.W. & Colby, D.R. (1993) Assessment of elemental contamination in estuarine and coastal environments based on geochemical and statistical modelling of sediments. *Marine Environmental Research* **36**, 237–66.

Kennicutt, M.C. II, Wade, T.L., Presley, B.J., Requejo, A.G., Brooks, J.M. & Denoux, G.J. (1994) Sediment contaminants in Casco Bay, Maine: inventories, sources, and potential for biological impact. *Environmental Science and Technology* **28**, 1–15.

Kerner, M. & Wallmann, K. (1992) Remobilisation events involving Cd and Zn from intertidal flat sediments in the Elbe Estuary during the tidal cycle. *Estuarine, Coastal and Shelf Science* **35**, 371–93.

Kitts, H.J., Millward, G.E., Morris, A.W. & Ebdon, L. (1994) Arsenic biogeochemistry in the Humber Estuary, U.K. *Estuarine, Coastal and Shelf Science* **39**, 157–72.

Li, Y.-H., Burkhardt, L. & Teraoka, H. (1984) Desorption and coagulation of trace elements during estuarine mixing. *Geochimica et Cosmochimica Acta* **48**, 1659–64.

McCain, B.B., Chan, S-L., Krahn, M.M., Brown, D.W., Myers, M.S., Landahl, J.T., Pierce, S., Clark, R.C. & Varanasi, U. (1992) Chemical contamination and associated fish diseases in San Diego Bay. *Environmental Science and Technology* **26**, 725–33.

Mackay, D.W. & Leatherland, T.M. (1976) Chemical processes in an estuary receiving major inputs of industrial and domestic wastes. In J.D. Burton & P.S. Liss (eds.), *Estuarine Chemistry.* Academic, London, 185–218.

Morris, A.W. (1986) Removal of trace metals in the very low salinity zone of the Tamar Estuary, England. *The Science of the Total Environment* **49**, 297–304.

Morris, A.W., Bale, A.J., Howland, R.J.M., Millward, G.E., Ackroyd, D.R., Loring, D.H. & Rantala, R.T.T. (1986) Sediment mobility and its contribution to trace metal cycling and retention in a macrotidal estuary. *Water Science and Technology* **18**, 111–19.

Muller, F.L.L., Tranter, M. & Balls, P.W. (1994) Distribution and transport of chemical constituents in the Clyde Estuary. *Estuarine, Coastal and Shelf Science* **39**, 105–26.

Ng, B., Turner, A., Tyler, A.O., Falconer, R.A. & Millward, G.E. (1996) Modelling contaminant geochemistry in estuaries. *Water Research* **30**, 63–74.

Pankow, J.F. & MacKenzie, S.W. (1991) Parameterising the equilibrium distribution of chemicals between the dissolved, solid particulate matter, and colloidal matter components in aqueous systems. *Environmental Science and Technology* **25**, 2046–53.

Plymsolve (1991) *An Estuarine Contaminant Simulator. User Manual.* Natural Environment Research Council, Swindon, 91 pp.

Preston, M.R. & Raymundo, C.C. (1993) The associations of linear alkyl benzenes with the bulk properties of sediments from the River Mersey Estuary. *Environmental Pollution* **81**, 7–13.

Robarbs, K. & Worsfold, P. (1991) Cadmium: toxicology and analysis. A review. *Analyst* **116**, 549–68.

Tanabe, S. (1988) PCB problems in the future: foresight from current knowledge. *Environmental Pollution* **50**, 5–28.

Turner, A. & Millward, G.E. (1994) The partitioning of trace metals in a macrotidal estuary. Implications for contaminant transport models. *Estuarine, Coastal and Shelf Science* **39**, 45–58.

Turner, A., Millward, G.E. & Morris, A.W. (1991) Particulate metals from five major North Sea estuaries. *Estuarine, Coastal and Shelf Science* **32**, 325–46.

Turner, A., Millward, G.E., Bale, A.J. & Morris, A.W. (1993) Application of the K_D concept to the study of trace metal removal and adsorption during estuarine mixing. *Estuarine, Coastal and Shelf Science* **36**, 1–13.

Tyler, A.O., Millward, G.E., Jones, P.H. & Turner, A. (1994) Polychlorinated dibenzo-para-dioxins and polychlorinated dibenzofurans in sediments from U.K. estuaries. *Estuarine, Coastal and Shelf Science* **39**, 1–13.

van Zoest, R. & van Eck, G.T.M. (1993) Historical input and behaviour of hexachlorobenzene, polychlorinated biphenyls and polycyclic aromatic hydrocarbons in two dated sediment cores from the Scheldt Estuary, SW Netherlands. *Marine Chemistry* **44**, 95–103.

Webster, J. & Ridgway, I. (1994) The application of the equilibrium partitioning approach for establishing quality criteria in two UK sea disposal and outfall sites. *Marine Pollution Bulletin* **28**, 653–61.

Zhou, J.L. & Rowland, S.J. (1997) The sorption of hydrophobic pyrethroid insecticides to estuarine particles: a compilation of recent research (this volume).

Zhou, J.L., Rowland, S.J. & Mantoura, R.F.C. (1995) Partition of synthetic pyrethroid insecticides between dissolved and particulate phases. *Water Research* **29**, 1023–31.

Zwolsman, J.J.G. & van Eck, G.T.M. (1993) Dissolved and particulate trace metal geochemistry in the Scheldt Estuary, S.W. Netherlands (water column and sediments). *Netherlands Journal of Aquatic Ecology* **27**, 287–300.

4

○ ○ ○ ○ ○ ○ ○ ○ ○ ○ ○ ○ ○ ○ ○ ○ ○ ○ ○ ○

A critical appraisal of the methodology used in studies of material flux between saltmarshes and coastal waters

K. Carpenter

4.1 Introduction

In the context of saltmarsh/coastal flux studies, a flux is the net exchange of a substance between these two ecosystems. The fluxes of several types of material, including nutrients, trace metals, herbicides and radionuclides, have been studied by researchers. Although this paper primarily appraises the methodology of nutrient flux estimations, many of the problems are also shared by researchers working on other types of material fluxes.

Since the first detrital flux studies conducted on Sapelo Island, Georgia, USA (Teal, 1962; Odum & de la Cruz, 1967), there has been a research interest in nutrient fluxes between saltmarshes and the adjacent coastal water. The major drive behind this work was to ascertain whether saltmarshes affect the concentration of dissolved inorganic nutrients and detritus (and hence the ecology) of coastal water. The link between saltmarsh processes and coastal productivity may be of relevance to coastal managers who wish to predict the impact of future saltmarsh loss (due to reclamation, erosion or drowning due to sea level rise) or possibly saltmarsh gain (if managed retreat becomes an extensively used strategy for coastal defence in rural areas).

The published flux results indicate that some saltmarshes appear to process nutrients in a very different way to others (Nixon, 1980; Carpenter, 1993). These discrepancies could be due to actual differences in nutrient budgeting caused by variations in saltmarsh characteristics. However, the disparity may be artificial, caused by differences in field methodology and the large uncertainties involved in the calculation of tidal and annual fluxes. It is difficult to differentiate between these two possibilities as error margins have never been reported for the published flux values. The total error on a flux value includes both methodology errors (e.g. uncertainties in the measurement of water discharge and nutrient concentration) and errors due to extrapolating instantaneous fluxes in space and time to produce tidal and annual fluxes. The resulting potential error margin may be larger than the calculated flux. Perhaps confidence limits are omitted by some researchers due to a natural reluctance to draw attention to this issue.

In this paper, I aim to draw attention to the large uncertainties which are involved in the estimation of annual nutrient flux values, using the various methodologies in current use. In particular, the paper highlights the differences in methodology (both field methods and during flux calculations) of flux studies to date, and how these would lead to major differences in the apparent flux even if none actually existed. It is hoped that by doing so, these errors may be reduced in future studies and that the data collected will be comparable. Only then may it be possible to establish how variability in saltmarsh characteristics such as plant species, temperature, and sediment type affect nutrient flux. This would bring researchers closer to the goal of developing a quantitative model that could apply to unsurveyed marshes to estimate their nutrient budgets and assess their influence on coastal ecology.

4.2 Differences in methodology which would cause variation in the calculated flux

A review paper published by Nixon in 1980 painstakingly brought together the research carried out in the previous 20 years on the role of saltmarshes in estuarine productivity and water chemistry. After considering the errors and assumptions of estimating annual flux using the direct method, he concluded that no research project had produced an accurate flux value. The major sampling problems included poor discharge estimates, incomplete knowledge of nutrient transport mechanisms and the unknown effects of storm tides on the total flux of materials (Wolaver *et al.*, 1983). Twelve years later this conclusion still stands as there has not been a significant improvement in the methods.

There is no standard method for determining the flux between saltmarshes and the coastal water. Differences in methodology arise during both the field work and the method of flux calculation. Table 4.1 lists the stages in flux measurement which have been tackled differently by saltmarsh researchers. Each of these stages will be discussed individually and I aim to point out that for a particular saltmarsh, the apparent flux of a substance would vary substantially depending on the set of methods chosen.

(a) Field technique chosen for flux measurement

There are four main field techniques which have been used to quantify nutrient fluxes. These are:

(a) community budgeting (e.g. Teal, 1962; Day *et al.*, 1973)
(b) direct tidal creek measurements (e.g. Axelrad, 1974; Valiela *et al.*, 1978; Roman, 1984; Dankers *et al.*, 1984; Whiting *et al.*, 1985; Azni, Abd. Aziz & Nedwell, 1986)
(c) flume studies (e.g. Wolaver *et al.*, 1983; Wolaver & Spurrier, 1988; Whiting *et al.*, 1989)
(d) diffusion chamber studies (e.g. Lee, 1979; Scudlark & Church, 1989; Chambers, 1992)

These methods are illustrated in Figure 4.1.

Methods (b), (c) and (d) usually involve sampling over a flood/ebb tidal cycle; the calculated tidal fluxes are subsequently extrapolated to give an estimate of the annual flux. Historically, method (a) was the first used. It is an indirect approach, whereby nutrient flux is derived from the difference between the annual production and aerobic decomposition. Teal used this method for his study published in 1962, which first cited saltmarshes as exporters of organic carbon. This method has the advantage of being long term, unlike tidal studies, which can only take snapshots of the flux during the sampled tides and may not be representative of the seasonal average. However, due to its indirectness, and the assumptions made (e.g. there was no anaerobic decomposition, nitrification or denitrification), this type of flux measurement was superseded by method (b).

Table 4.1 Stages in the methodology in which differences would change the calculated flux value

Field techniques
- (a) Technique chosen for flux measurement
- (b) Position of sampling station in marsh system
- (c) Water sampling strategy
- (d) Accuracy of discharge measurement
- (e) Frequency of discharge and concentration measurements
- (f) Frequency of tidal monitoring

Flux calculations
- (g) Tidal flux calculations:
  - (i) Combination of discharge and nutrient concentration to obtain flux
  - (ii) Treatment of tidal inequality
- (h) Annual flux calculations:
  Method of extrapolating results from the sampled tides to give annual flux

Method (b) is a direct method that measures the amount of nutrient flowing into and out of the marsh through a closed tidal creek system during a flood/ebb cycle. To calculate the total number of moles of a nutrient species entering the marsh system during the flooding portion of the tide, the concentration of this nutrient and the rate of water discharge throughout the flood tide are multiplied together. The number of moles leaving the marsh with the ebbing tide is similarly calculated and the flux is the difference between the estimated flood and ebb totals. The value of the results from this method are greatly related to the accuracy to which water discharge can be measured (see section (e)). This method integrates the results of all the nutrient transforming processes occurring in the various subsystems of the saltmarsh (e.g. creek water column, sediment, vegetated surface and creek bank) and thus gives the net flux value for the marsh as a whole, unlike methods (c) and (d).

The flume technique (c) has been used by Wolaver and his colleagues to measure marsh surface processes on the marshes of North Inlet, South Carolina, USA. It can only be used for tides that are large enough

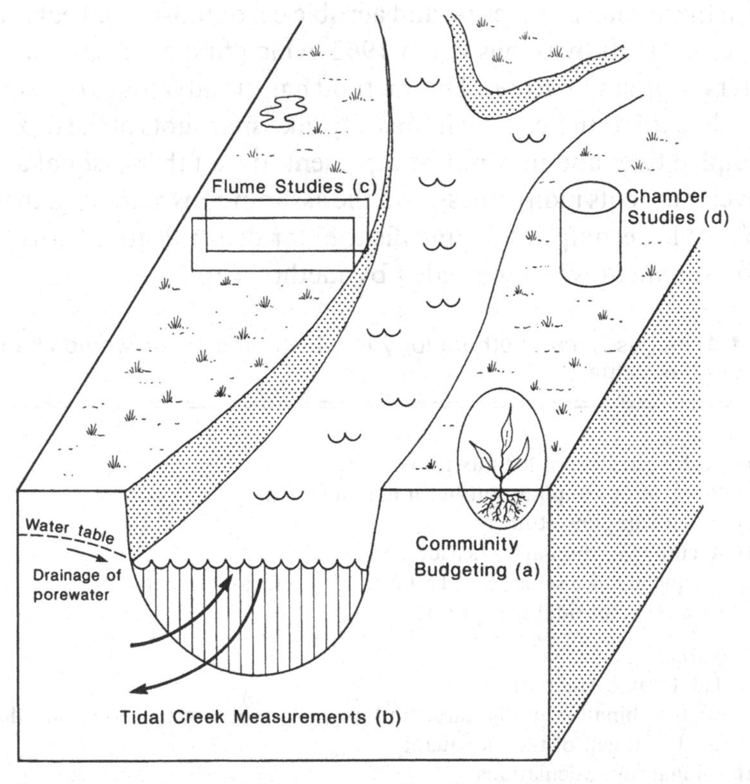

Figure 4.1 Schematic representation of four main techniques used to quantify nutrient fluxes between saltmarshes and the coastal water.

to spill over the banks and onto the marsh surface. The flume consists of two walls of plexiglass fencing pushed partly into the sediment, parallel to the flow of overtopping water, stretching from the creek or mudflat into the interior of the marsh. To estimate the nutrient flux from this subsystem, water samples are taken from the end of the flume nearest the creek, throughout the portion of the tidal cycle when water is flowing over the marsh surface. The flux is again estimated by combining nutrient and discharge data, but the flow is calculated from the change in volume of the water within the flume per unit time instead of using velocity meters. Water volume estimates are made using water level elevation and the topography of the marsh surface within the flume (Wolaver *et al.*, 1983).

It is important to realise that the flume method should not be used to estimate the overall flux from the saltmarsh. It is only suitable for assessing the importance of surface processes, such as diffusion of substances between pore-water and inundating tidal water, uptake by vegetation and wash off processes. The results can not be extrapolated to determine the total flux between the saltmarsh and the coastal water, as the nutrient transforming processes associated with the marsh surface only play a small role in determining the overall nutrient flux (see Figure 4.2).

The fourth method (d) which has been used to measure saltmarsh nutrient fluxes is the diffusion chamber technique. The basic apparatus consists of a perspex cylinder which is partly pushed into the surface of the marsh and filled with water. Periodically water samples are removed from the chamber and analysed for nutrient species. It would seem logical to use creek water taken at high tide, so that an increase or decrease in the nutrient concentration with time would indicate whether the marsh surface was a sink or source under natural concentration gradients.

Chambers (1992) resolved a major problem associated with the earlier chamber studies. These had an artificial flooding regime, where the depth of the water above the marsh surface remained constant throughout the tidal cycle. He managed to produce a natural flooding regime in the chamber using an ingeniously simple design, involving a collapsible reservoir which floods and drains the chamber with water. Water movement between the reservoir and the chamber is controlled by the pressure of the water on the marsh surface. As the water that floods the chamber is isolated and separated from the tidal waters, this experimental design is ideal to investigate how chemical and biological variables affect nutrient cycling. Another advantage of the chamber method is that the flux from small areas can be determined with confidence; this would be useful for studying the effect of different plant species, for example. Furthermore, by comparing the predicted flux values (calculated from pore-water concentration gradients, using a diffusion model) with the actual flux rates (derived from flux chamber work), the relative importance of surface and subsurface processes in regulating sedimentary nutrient fluxes can be evaluated (Scudlark & Church, 1989).

Like the flume method, fluxes derived from chamber experiments do not integrate all marsh processes, and only measure changes in nutrient concentration caused by diffusion or uptake during over bank tides; the technique may also perturb the system. It is also inappropriate to use this technique to estimate annual nurient fluxes, via the marsh surface, if the calculated fluxes are based on constant inundation of the marsh surface (as were those of Scudlark & Church, 1989) and not the periodic submergence which actually occurs. It would be relatively easy to correct for this error, if surface water level had been continually monitored at the site, by calculating the periods of inundation from the tidal records and details of the marsh topography. However, it should be noted this extrapolation is another source of uncertainty.

To summarise, flume and chamber studies are helpful for investigating the relative importance of surface diffusion, wash off and plant uptake/release from aerial shoots, when used in conjunction with direct tidal creek flux measurements. However, the results from these experiments should not be extrapolated to give a flux value for the marsh as a whole as they do not account for sediment drainage, water column and creek bank processes.

Figure 4.2 illustrates why the direct tidal creek method is best for determining whether a saltmarsh is a source or a sink of nutrients to the coastal water. The saltmarsh nitrogen cycle has been used as an example to indicate which of the four main techniques, outlined earlier, accounts for each of the sources and sinks of nitrogen. Although the direct creek method treats the saltmarsh as a black box (so that the sites at which the nutrient transforming processes occur cannot be identified without further work), it is the only method which integrates the nutrient transforming processes in all the different subsystems. The other methods only account for some of these processes.

The difference the field methodology makes to the estimation of the nutrient flux between a saltmarsh and coastal water can be illustrated using studies conducted on North Inlet marsh, South Carolina. Wolaver & Spurrier (1988), used the flume technique to estimate the budget for a whole marsh. They concluded that North Inlet marsh imported both phosphate and particulate phosphorus. However, an earlier direct tidal study in the same marsh by Whiting *et al.* (1985) reported an export of these two nutrient species. Although Whiting's study was confined to eight tidal cycles in a single month (May), and thus cannot be considered as an annual budget, it still shows that the integrated approach was not in agreement with the flume study as the latter found the marsh to be an importer in all seasons. Furthermore, all the other American flux studies apart from those using chambers or flumes, found saltmarshes to be net sources of phosphate. Thus, from the field studies, it does appear that flume or chamber studies do not reflect the total flux from the marsh, as predicted earlier from the fundamental flaws in the approach.

It is not surprising that flume chamber techniques indicate a different flux direction for phosphate. It is well known from fertilisation experiments (e.g. Tyler, 1967; Marshall, 1970; Patrick & Delaune, 1976; Valiela & Teal, 1976) that primary production on saltmarshes is nutrient limited, so nutrients will be taken out of the overlying water and converted into biomass, the majority of which will be below ground. After death, some of

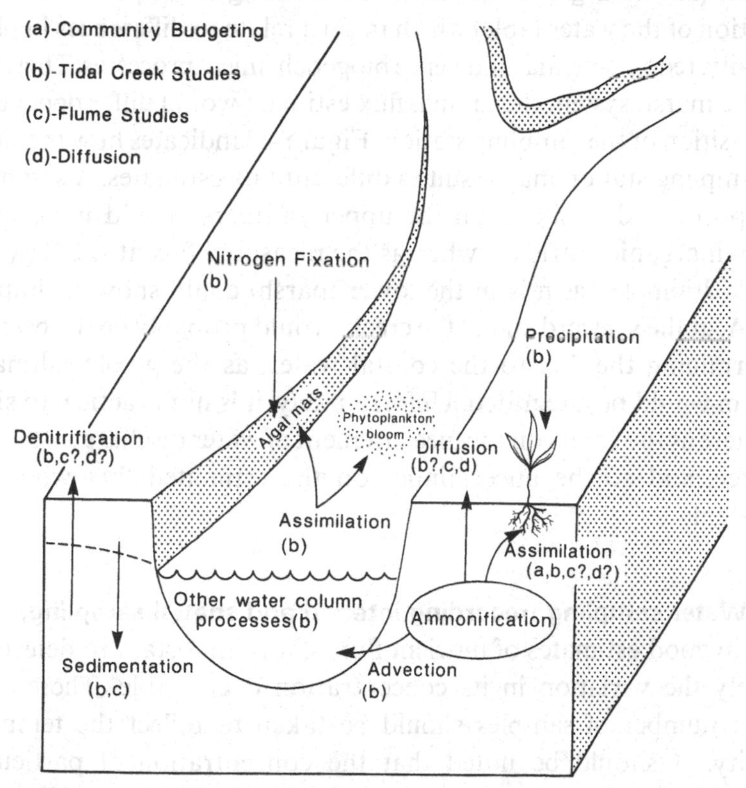

Figure 4.2 Diagram to illustrate the sources and sinks of fixed nitrogen in a saltmarsh and to indicate which of these processes are taken into acount by each of the alternative techniques for the estimation of 'saltmarsh/coastal water' flux.

this biomass, plus organic matter from marsh sedimentation, will be remineralised in the sediment to release the inorganic and organic nutrients to the pore-waters. Drainage of saltmarsh groundwater into the creek, followed by tidal flushing, may export these nutrients from the marsh. Therefore, a marsh could be a source of dissolved inorganic nutrients even if surface processes suggest it to be a sink.

(b) Position of sampling station in marsh system

The next methodology difference which could cause disparity of flux results, even if researchers were working on the same marsh, is the position of the sampling station. Younger, lower marshes seem to function differently to the older, upper marshes; there is evidence that the former are net importers while the latter are net exporters (Boorman *et al.*, 1994; Dame, 1991). The variation in the way nutrients are cycled within these two areas of the marsh probably relates primarily to hydrological differences (i.e. the degree of drainage and flushing of the pore-water and the position of the water table) which in turn relate to differences in plant community, redox potential and hence biogeochemical processes. Therefore within the marsh system the annual flux estimate would differ depending on the position of the sampling station. Figure 4.3 indicates how the siting of the sampling station may result in different flux estimates. A sampling station positioned at site C on the upper saltmarsh could indicate an export of inorganic nutrients, whereas the measured flux at site B (where the creek drainage basin is in the lower marsh) could show an import. Station A, at the seaward end of the creek, would probably be the best site for determining the flux to the coastal water, as the whole saltmarsh ecosystem would be integrated. Unfortunately, it is not practical to site a station here as the errors on the measurement of water discharge would be very large, and so the uncertainties on the estimated flux would be unacceptable.

(c) Water sampling (recording interval and spatial sampling)

To obtain good estimates of nutrient fluxes, it is important to determine accurately the variation in its concentration over a tide. Therefore a sufficient number of samples should be taken to reflect the temporal variability. It should be noted that the concentration of particulate material is prone to greater short-term fluctuations than dissolved constituents and so the former requires more intensive sampling than the latter. In addition to temporal variability, spatial variability needs also to be considered. The water in large creeks is not necessarily well mixed. To determine an adequate sampling regime an initial experiment should be conducted to determine the degree of spatial variation in nutrient concentration across the creek cross-section. Again, the spatial variation is greater for particulate than dissolved materials as suspended solid concentration varies strongly with depth, with more sediment moving close to the bed than at the surface.

To ensure that the water samples maintain the same nutrient concentration until they are analysed, the samples should be treated (e.g. by filtering, addition of biocide, freezing) as soon as they are collected.

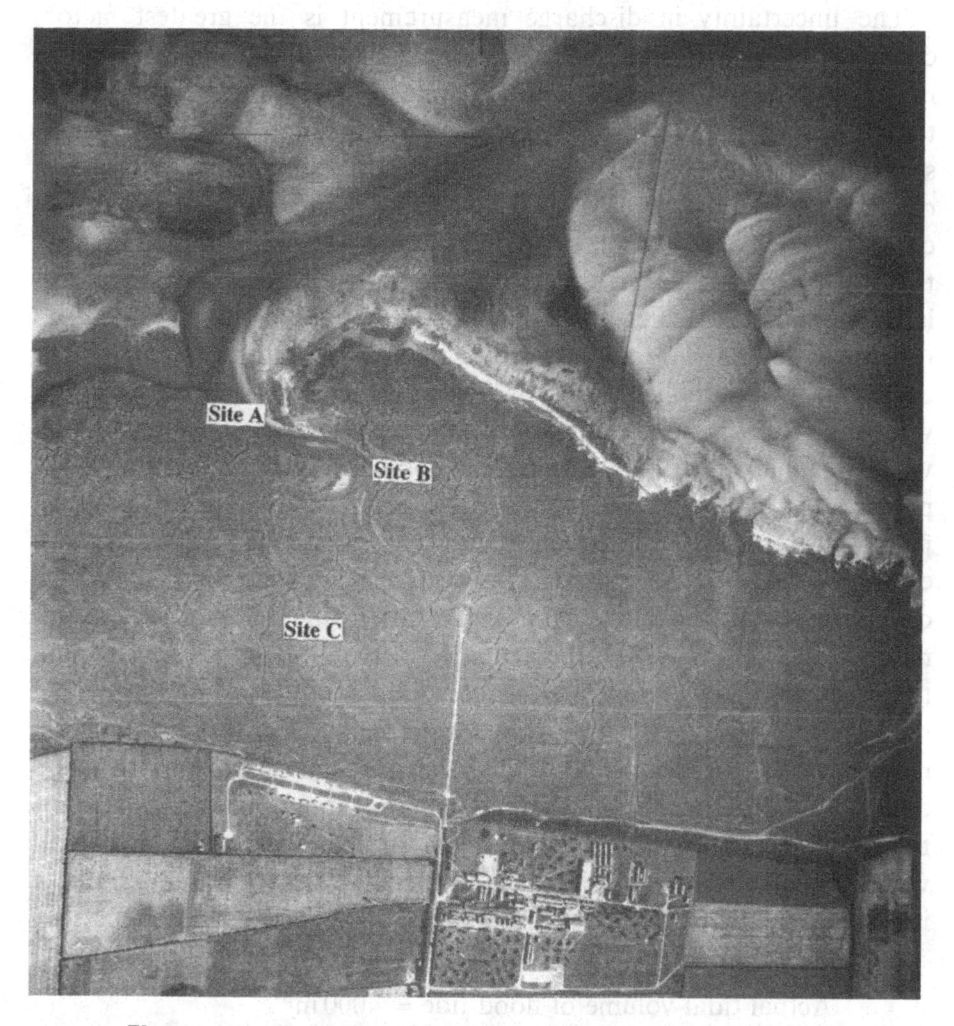

Figure 4.3 Illustration of how the positioning of sampling stations may influence the estimated flux, due to the different nature of the drainage basins.

Failure to adequately preserve water samples results will compromise the estimated flux values.

(d) Accuracy of water discharge estimates

The main difficulty with direct tidal creek measurements is obtaining reliable water velocity/discharge data. The hydrology of tidal creeks is different from fluvial channels due to tidal flow which causes rapid change in water level and the rate of discharge (making accurate gauging by velocity meters extremely difficult). Furthermore, the bidirectional nature of water flow in creeks renders gauging by weirs inappropriate.

The uncertainty in discharge measurement is the greatest factor contributing to the error on the calculated tidal nutrient flux. Velocity meters have been universally used to gauge flows; most tide studies have taken velocity readings from only one or two points across the cross section of the creek and used this value to estimate instantaneous discharge across the whole creek. The hydrodynamics of tidal creeks are complex (Kjerfve & Proehl, 1979) and are not well suited to gauging by this method, hence the error on an instantaneous discharge estimate is large. Roman (1984) calculated hourly instantaneous discharge values (i.e. the rate of discharge through the entire creek cross section) for three tidal cycles using 18 sampling positions. The average cross-sectional velocity was found for each hourly interval and these data were regressed with the velocity values from the 18 sampling positions. The station that best predicted the average cross-sectional velocity was identified and this position alone was subsequently used to gauge discharge. Roman estimated that the average error between the calculated instantaneous discharges using the single sampling position and the dense array of meters was 11%. However, the actual error would be much larger than this as even the densely metered value would still only approximate the real instantaneous discharge. Furthermore, with only 12 discharge measurements for the whole tidal cycle, irregular surges in discharge could be missed, thus the random error on the total discharge volume for the flood and ebb positions of the tide could be substantial. This error would have an effect on the value and even the direction of the flux; as illustrated below. Assume, for example,

Actual tidal volume of flood tide = $4000\,m^3$

Actual tidal volume of ebb tide = $4500\,m^3$

Actual total NH_4^+ in flood tide = 30 moles
Actual total NH_4^+ in ebb tide = 40 moles
Therefore, **actual flux = − 10 moles**

If the measured flood discharge was 30% higher and the ebb 10% lower than the actual, and the variation in concentration through the tide was monitored accurately, the results would be:

Estimated flood discharge = 5200 m^3
Estimated tidal volume in ebb = 4050 m^3
Calculated NH_4^+ in flood tide = 39 moles
Calculated NH_4^+ in ebb tide = 36 moles
Therefore, **estimated flux = + 3 moles**

The hypothetical calculation above illustrates how crucial accurate discharge data is for accurate flux estimates and the necessity for an accurate gauging method.

An improvement in the methodology to measure flows through tidal creeks would be the single most efficient way of reducing the uncertainty on the calculated flux value. In my research on Stiffkey marsh, North Norfolk, UK, I used an alternative method of gauging creek flow that was developed by David Sargent (Sargent, 1981, 1982a, b; Carpenter, 1993). This method, known as the integrating rising air float technique, is ideally suited to overcoming the difficulties in gauging tidal creek flows, which velocity meters fail to overcome. The technique works on the principle that horizontal displacement of a buoyant float rising from the bed to the surface, is proportional to the unit discharge. Air is pushed through a flexible pipe, which is pulled straight and lies along the banks and bed of the creek, using the air pressure from an aqualung type cylinder (see Figure 4.4). The air exits the pipe from nozzles which occur at 0.5 m intervals along the pipe. As the bubbles rise they are displaced laterally by the current. The major advantage of this technique is that the bubbles integrate the varying flow velocities through the water column; velocity meter readings, at various depths down a vertical, only sample this velocity profile. Furthermore, instantaneous discharge across the whole creek cross section can be gauged by this method as frequently as desired by recording the positions of the surfacing bubbles across the creek with photography and deducing the horizontal area between the air pipe and the bubble envelope (area A in

Figure 4.5) using photogrammetry. Discharge through the creek is calculated by multiplying this area with the rise velocity of the bubble. Photographs could be taken every few minutes during periods of the tidal cycle experiencing the greatest changes in discharge; hence this method could pick up discharge pulses which velocity meters may miss if the frequency at which instantaneous discharge can be calculated is too low.

The instantaneous rising air float technique is not only inexpensive to make and involves little field labour, it is also measures water discharge through creeks more accurately than velocity meters. Sargent (1981) calibrated the rising air float technique against a gauging weir and found the average discrepancy was 1.3% (although the range was – 6 to + 10%). This is considerably lower than the error on the creek water discharge rates using the velocity meter method. Hence the integrating rising air float technique seems superior to velocity meters and should be considered as a preferable gauging option in future saltmarsh material flux studies.

Figure 4.4 The equipment used in the integrating rising air float technique.

Figure 4.5 Photograph of the emerging air bubbles, from which the rate of water discharge can be calculated. Discharge = area A × bubble rise velocity. Area A is calculated using photogrammetry. PF is the principal focus.

(e) Frequency of discharge and concentration measurements

The last major difference in the way researchers have conducted the field work of tidal flux studies is in the frequency of recording water discharge and nutrient concentration: ranging from 10 minutes to 1 hour, depending on the position in the tidal cycle (Carpenter, 1993); every 30 minutes (e.g. Boorman *et al.*, 1994) to every 1.5 hours (Whiting *et al.*, 1985). The error of extrapolating the data to estimate the total flow during the tidal cycle increases with time between measurements. To establish the optimal sampling frequency, it would be a good idea to take water samples and discharge measurements every five minutes and use this data set to investigate the difference that various sampling frequencies have on the calculated flux value.

(f) Frequency of tidal monitoring

Nutrient flux has been found to differ both seasonally and with tidal range (spring–neap cycle). The seasonal variation is illustrated by data collected during my PhD research of nutrient fluxes between Stiffkey upper saltmarsh, on the North Norfolk coast, and the North Sea. Figure 4.6 illustrates the seasonal differences between nutrient concentrations in the North Sea with those in the water seeping from the saltmarsh sediments. The bars represent the concentration at the end of the ebb tide just before the flooding water of the following tide reached the sampling station (when the water flow is fed by groundwater seepage from the saturated saltmarsh sediment). The points joined together by a line are the concentrations at high water (which most closely resembles the character of coastal sea water). By comparing the nutrient concentrations at these two points in the tidal cycle, it is possible to establish whether the net effect of the nutrient transforming processes in the sediment is uptake or release. Thus when the bar is larger than the line, the sediment–creek bank system is a source of the nutrient and conversely a higher line than the bar denotes the system is a net sink. It is clear that the size of the flux varies considerably, and in the case of phosphate and nitrate the direction of the flux also differs seasonally.

In order to get a good estimate of annual flux, tidal monitoring should be as frequent as possible to reduce the errors associated with interpolation and extrapolation. Ideally, several tides, covering the range of tidal heights should be sampled each month. This is particularly important for sediment flux studies as suspended sediment concentration varies with tidal range with the highest concentrations occurring at spring tides due to the associated large tidal velocities. Furthermore, attempts should be

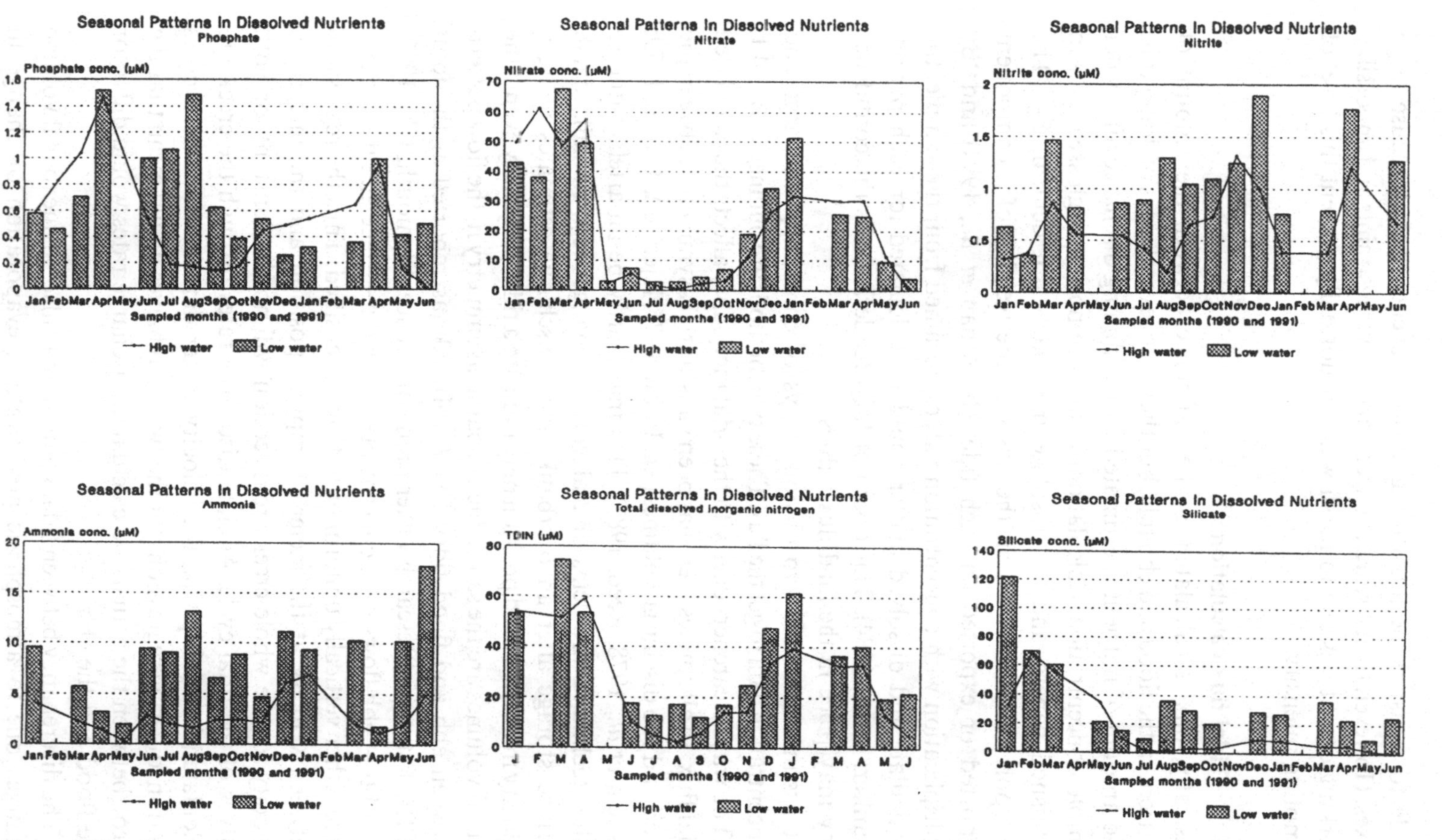

Figure 4.6 Seasonality in nutrient concentration in the creek water at high and low tide, Stiffkey saltmarsh, North Norfolk UK.

made to measure the flux during episodic events likely to cause large fluxes. However, tight research budgets, manpower restraints and impossible working conditions due to harsh weather normally prevent this level of sampling intensity.

(g) Tidal flux calculations

Basically, flux is calculated as the difference between the amount of nutrient in the flood and ebb tide, but there are several ways of doing this. The amount of nutrient is estimated by combining discharge and water sample concentration. Calculation details are normally glossed over in published papers, and the absolute method is not defined, e.g. 'The transport was calculated on the basis of the weights of a component imported or exported on each tide' (Boorman *et al.*, 1994) and 'By multiplication with concentration values obtained from the sample, the particulate and dissolved matter could be calculated for each period' (Dankers *et al.*, 1984). This does not help readers assess the confidence they may place in the final flux values.

I assume there are two obvious ways of combining discharge and concentration data to get the flux. These two methods are outlined in Box 4.1.

The way researchers deal with the tidal inequality affects the calculated tidal flux. In all studies there has been substantial asymmetry between the flood and ebb discharge volumes (e.g. Boon, 1975; Kjerfve & Proehl, 1979; Valiela *et al.*, 1978; Reed, 1987). This may partly be an artefact due to either technique problems or real volume changes (e.g. rainfall, marsh surface storage after an overbank tide). Discharge estimates using a velocity meter in a fixed position may indicate a tidal inequality even if the actual volumes are the same, due to spatial asymmetry in the flow pattern between ebb and flood tides. Jay & Musiak (as reported at the joint ECSA/Estuarine Research Federation conference, Plymouth, UK, 1992) found that ebb flows were concentrated near the surface and flood flows were either vertically uniform or bottom intensified. Thus the regression between velocity at the meter's sampling position and the integrated velocity for the whole creek cross section will not be constant for both halves of the tidal cycle. As tidal discharge rates through the creek are calculated by multiplying the velocity meter data by the wetted cross section, differing regressions for the two tidal halves would mean that the error between the actual and calculated discharge rates would differ for the flood and the ebb.

The discrepancy between tidal volumes may also be due to real volume changes due to atmospheric precipitation, evaporation or change in

Box 4.1 Two possible ways of combining discharge[1] and concentration[2] data to obtain flux values

(1) By interpolation of best fit lines through Q and C data

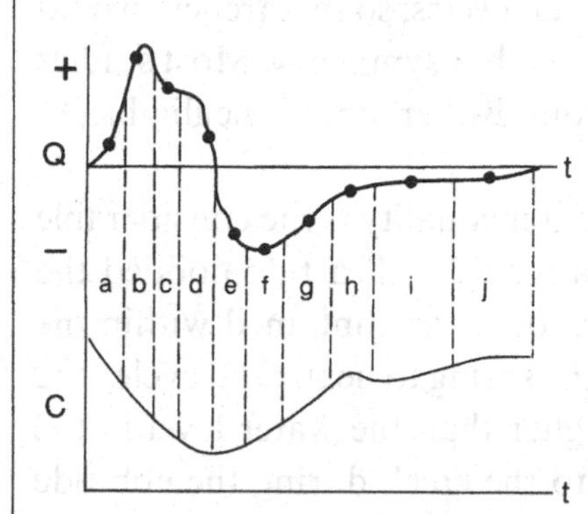

$$\text{Total no. of moles in flood } (A) = \sum_{a}^{d} (\bar{Q}\cdot t)\cdot\bar{c}$$

$$\text{Total no. of moles in ebb } (B) = \sum_{e}^{j} (\bar{Q}\cdot t)\cdot\bar{c}$$

Tidal flux $= A - B$

(2) Via the flow rate of substance (mol s^{-1}) from the multiplication of Q and C data which were recorded simultaneously

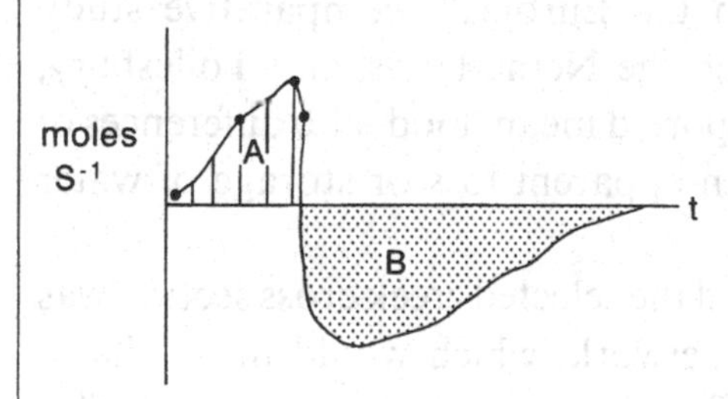

Tidal flux = Area A – Area B

storage. Precipitation effects are usually insignificant due to the small ratio between precipitation volume and tidal input. Land-derived ground-water input could potentially cause a significantly inflated ebb volume (e.g. in the Sapelo Island marshes), especially in sandy marshes backed onto land of high elevation. However, it is unlikely that this is the dominant cause of the tidal discrepancy as it does not explain the many cases which exhibit significantly larger flood than ebb volumes. Furthermore, large inputs of fresh water would easily have been identified by researchers from changes in salinity over measured tidal cycles, so researchers would have noticed if groundwater input explained the asymmetry. Most studies have selected sites where land-derived groundwater input is negligible, to simplify their calculations.

One of the most likely causes of the tidal inequality is the considerable change in stage height between consecutive tides. If a tide flooded the surface of the marsh, but the subsequent one was contained within the banks of the creek (as may occur during the spring to neap tidal cycle), the water table in the sediment could be higher than the water level in the creek. Thus more water would drain into the creek during the ebb tide than infiltrates through the creek banks during the flood tide. Hence the ebb volume would be larger than the flood. During the neap to spring cycle, the reverse situation exists and the flood could be bigger than the ebb. The first 'overbank' tide of the cycle would theoretically show the biggest discrepancy between flood and ebb volumes because flood water would remain on the marsh surface due to the replenishment of pans and infiltration into the unsaturated surface sediment. If the described events were the major cause of observed tidal asymmetry, one would expect that randomly sampled tides would be equally split between ebb dominated and flood dominated tides. This does not seem to be the case in at least some studies, the three marshes in the European comparative study (Mont St Michel, France; De Slufter, the Netherlands; and Tollesbury, UK) which measured creek flows, reported mean flood–ebb differences of − 13%, − 11%, and − 16%, i.e. an apparent loss or storage of water (Boorman *et al.*, 1994).

Tidal asymmetry could also occur if the selected creek cross section was not situated within a closed creek network, which would mean that a parcel of water could exit from a different place to where it entered. This could occur if two creek systems were connected, or could exchange water during an overtopping tide. The proportion of water entering and leaving the marsh from the two mouths would depend on wind direction and tidal flow patterns.

Some researchers have ignored inequality in the flood and ebb volumes completely (e.g. Dankers *et al.*, 1984), whilst others have compensated for the tidal discrepancy by multiplying the total amount of nutrient entering the marsh on the flooding tide by the ratio of the flood/ebb volume (e.g. Valiela *et al.*, 1978). The way researchers treat the asymmetry obviously has a great effect on the calculated flux value, so which is the best approach to adopt? It would be incorrect to equalise an ebb flow, which has been reduced in volume by evaporation, to the preceding flood, as this could falsely show an export of material when none existed. On the other hand it would be erroneous to ignore the inequality of a large tide which had greater flood than ebb, due to recharge and surface ponding. The calculated results most likely show a marsh input of material even though the nutrients may still exist unchanged in the ponded water and may later return to the creek via advection. It would be even worse not to equalise tides that were unbalanced due to transference of water between interconnected creek systems. In terms of nutrient cycling and determining the direction of the flux, it does not matter that some of the ebb water entered the marsh via a different creek mouth, as that water package has experienced the same marsh processes as water that entered from the main creek mouth. However, it is possible that the magnitude of the flux per meter cubed of tidal water is altered if the secondary package had spent a different length of time on the marsh or covered less surface area. Thus if the tidal inequality was due to the interconnection between two or more creek systems it would be correct to equalise the volumes. Therefore, although there are errors associated with both ignoring and compensating for the flow imbalance, it seems better to equalise the flows unless major precipitation or evaporation has taken place.

(h) Annual flux calculations

The confidence that can be placed in annual flux values is even lower than for individual tidal cycles. Previous research has concentrated on the determination of a value for annual flux, almost all papers reporting this value despite the uncertainty connected with it. Less attention seems to be paid to seasonal variation in the flux value, which seems odd as the direction of the flux may change throughout the year but the net annual flux may be insignificant, thus masking the true impact the saltmarsh has on the ecology of the adjoining coastal water. For example, a saltmarsh may be a sink of phosphate in the winter and a source in the summer (Asjes, presentation at INTECOL's IV International Wetlands conference, 1992, relating to a Dutch saltmarsh; Carpenter, 1993). The annual flux of phosphate from Asjes's study was negligible: if this value was presented alone (as in the style of Nixon, 1980; Dankers *et al.*, 1984, for example) one would conclude that phosphorus cycling in the marsh is insignificant. However, the seasonal fluxes are potentially very important to coastal production as phosphate is released at a time when the concentration in coastal water is at its minimum.

Calculated nutrient fluxes between Stiffkey saltmarsh and the North Sea, from my PhD research (Carpenter, 1993) are given in Tables 4.2 to 4.4. These three tables express the results in different ways which suit different purposes. Table 4.2 expresses the semi-diurnal tidal flux in moles. This total value is useful for evaluating the impact the saltmarsh has on nutrient concentration in the coastal water. Table 4.3 describes the fluxes in volumetric terms, i.e. the flux per cubic metre of the flood tide. This is calculated by dividing the semi-diurnal tidal nutrient flux by the averaged tidal volume. This form of expression is very useful for making comparisons with other research. The total tidal flux can not be used to compare saltmarshes as the value depends on both the drainage area and the tidal volume. Table 4.4 expresses the fluxes as a percentage of the total number of moles in the flood tide (after the quantity in the flood had been corrected for tidal volume inequalities). This gives an idea of the size of the flux in comparison to the amount flowing through the creek during tidal flushing and also the degree of confidence which can be placed in the direction and absolute value of the calculated flux. The values depicted in bold type indicate the fluxes which were larger than the uncertainty. The flux results show that there is seasonal variation in all the measured nutrient species, which is not appreciated if just the averaged annual flux is reported.

Table 4.2 Nutrient fluxes (moles) for measured semi-diurnal tidal cycles, Stiffkey Saltmarsh, North Norfolk, UK. A negative sign denotes the saltmarsh as a source of nutrients to the offshore waters

		Flux (moles)				
Date		NO_2^-	NO_3^-	NH_4^+	PO_4^{3-}	SiO_4^{3-}
12 Aug. 1990		– 2.8	1.9	– 1.94	– 1.99	– 50.7
12 Oct. 1990		– 0.88	– 0.19	– 7.09	0.1	3.17
8 Nov. 1990		0.21	– 1.9	1.5	– 0.027	–
22 Jan. 1991		0.24	– 6.5	– 1.79	0.035	– 7.69
22 Mar. 1991		0.03	0.21	8.16	0.71	20.1
20 Apr. 1991		– 0.42	– 6.7	6.83	0.16	10.7
19 May 1991		–	10.7	2.41	0.13	4.76
14 Jun. 1991	Spring	– 0.001	2.7	4.5	– 0.01	– 12.5
15 Jun. 1991	to	– 0.60	– 10.9	– 2.67	– 4.45	– 58.9
17 Jun. 1991	Neap	–	7.2	2.5	– 0.38	– 24.7
19 Jun. 1991		–	5.3	5.04	– 0.23	– 4.3
21 Jun. 1991 ↓		– 0.36	2.2	– 1.02	– 0.11	– 2.8

To my knowledge, there are four major methods of calculating the annual flux from tidal flux estimates. All four methods suffer from assumptions and huge extrapolations of the data set, to the extent that I did not attempt to calculate an annual flux for my study as I did not think the resulting figure would be meaningful. Instead I compared the tidal fluxes which were measured monthly during the same point of the spring–neap tidal cycle to identify seasonal changes in nutrient fluxes.

It has not been possible to decipher the methods of annual flux calculation in detail, as they are not usually reported in published

Table 4.3 Volumetric nutrient fluxes (flux per cubic metre of tidal volume), Stiffkey Saltmarsh, North Norfolk, UK

	Volumetric flux (μmoles m^{-3} of tidal volume)				
Date	NO_2^-	NO_3^-	NH_3	PO_4^{3-}	SiO_4^{3-}
12 Aug. 1990	– 282	186	– 194	– 199	– 5070
12 Oct. 1990	– 152	– 33	– 1226	17	548
8 Nov. 1990	121	– 1095	846	– 15	–
22 Jan. 1991	82	– 2182	– 601	12	– 2581
22 Mar. 1991	5.2	3.8	1471	128	3677
20 Apr. 1991	– 92	– 1472	1493	35	4461
19 May 1991	–	1803	406	27	801
14 Jun. 1991	– 0.6	163	271	– 0.6	– 749
15 Jun. 1991	– 17	– 307	– 75	– 125	– 1655
17 Jun. 1991	–	652	226	– 34	– 2229
19 Jun. 1991	–	936	896	41	– 760
21 Jun. 1991	– 210	1302	– 595	– 64	– 1535

Table 4.4 Nutrient fluxes as a percentage of the total number of moles in the flood tide, Stiffkey Saltmarsh, North Norfolk, UK

	Percentage flux (%)				
Date	NO_2^-	NO_3^-	NH_3	PO_4^{3-}	SiO_4^{3-}
12 Aug. 1990	– **140.0**	**12.4**	– 6.7	– **31.0**	– 61.8
12 Oct. 1990	– 3.5	– 3.8	– **23.0**	6.9	5.5
8 Nov. 1990	9.0	– 9.1	– **21.7**	– 3.4	–
22 Jan. 1991	**12.0**	– 5.3	– 1.8	**18.9**	– 12.2
22 Mar. 1991	0.9	0.1	**22.9**	**16.8**	20.9
20 Apr. 1991	– 6.5	– 4.6	**40.7**	2.9	13.5
19 May 1991	–	**13.9**	11.4	9.7	16.0
14 Jun. 1991	– 0.2	**43.0**	**17.9**	5.7	– 16.8
15 Jun. 1991	– **30.6**	**48.9**	– 3.5	– **42.0**	– 133
17 Jun. 1991	–	**27.1**	7.5	– **14.2**	– 46.1
19 Jun. 1991	– **22.6**	**21.0**	**27.1**	– **14.1**	– 10.3

literature, but the general approaches are given below. It is possible that each method would give a different flux value given the same data set.

(i) The simplest method averaged the fluxes of the tides sampled throughout the year and multiplied this value by 706, i.e. the total number of tides per annum (Wolaver *et al.*, 1983; Whiting *et al.*, 1989). In these two studies, tidal cycles were sampled once every 2–4 weeks, selecting tides to cover all ranges and times of day. It would be very difficult to investigate seasonal nutrient cycles using this method as the flux is affected by both season and tidal range (Wolaver & Spurrier, 1988), i.e. if a spring tide was sampled in summer and a neap tide in autumn, seasonal variation would be masked by tidal height effects.

(ii) Dankers *et al.* (1984) grouped their sampled tides into bands governed by the elevation of high water. For each group the mean flux was calculated and multiplied by the number of tides in the year within that range. The resulting products were summed to give the annual flux. This method assumed that tidal height, rather than seasonal factors, is the dominant variable controlling the flux.

(iii) Asjes (pers. com., 1992) took storm tides into account. He managed to measure the tidal flux during a storm tide (this has rarely been achieved) and assumed that all storm tides throughout the year had the same flux. These fluxes were included in the annual flux estimation.

(iv) Arguably the most accurate way of calculating annual flux would be through the use of a regression model. By knowing how the key variables behave throughout the year the fluxes for the unsampled tides could be estimated. However, saltmarsh science has not yet reached the stage at which this could be possible. Wolaver & Spurrier (1988) made a contribution towards this aim when they correlated their flume derived flux results, from a South Carolina saltmarsh, with pertinent chemical, biological and physical parameters, to find the most significant factors influencing phosphate flux. They concluded that the major controlling variables, which individually explained up to 10% of the flux, included tide height, water temperature, salinity, wind and rain. However, they did not attempt to produce a multiple regression equation for these data so that phosphate flux between an unstudied marsh and the coastal water could be estimated.

4.3 Summary

(1) Nutrient fluxes have been obtained using a range of field techniques and methods to calculate tidal and annual fluxes. The methodology

chosen affects the value of the estimated flux. In order to compare how various saltmarshes process nutrients, research institutes should coordinate studies using the same methodology.

(2) Four field techniques have been used to estimate the import/export of nutrients between saltmarshes and the adjacent coastal water: community budgeting, direct creek studies, flume studies and diffusion chamber studies. The direct tidal creek method is the only one which takes account of the processes in all of the saltmarsh subsystems and hence appears to be the best technique, provided the uncertainties in water discharge are minimised.

(3) In order to determine whether the differences in nutrient fluxes from saltmarshes around the world are due to real differences in nutrient budgeting as opposed to uncertainties in the calculated value, confidence limits should be assigned to published flux values.

(4) The uncertainties in calculated flux values are large due to methodology errors and errors associated with extrapolating the collected data in space and time. A major problem with saltmarsh/coastal water flux studies is that the flux is often smaller than the error margin, hence one is not even sure of the direction of the flux. The error associated with an annual flux estimate is much higher than the tidal flux.

(5) Past researchers have concentrated on producing an annual flux value, nearly all reporting this value (despite the huge uncertainties) rather than drawing attention to seasonal variations in flux. These will be masked by an annual average, but have a significant effect on the ecology of the adjacent water body.

(6) The confidence that can be placed in material flux studies relies heavily on the accuracy of the method for measuring water flow through the creek. Sargent's integrating rising air float technique has a significantly lower uncertainty on the calculated instantaneous discharge through tidal creeks than velocity meters. The adoption of this technique could make an important contribution to the accuracy of future tidal creek material flux studies.

(7) The validity of nutrient flux studies as a method to establish whether saltmarsh processes have a significant effect on the productivity of coastal water should be questioned. Unless uncertainties are reduced by improvements in methodology, and an intensive sampling regime undertaken (sampling a semi-diurnal tidal cycle every few days), creek nutrient flux studies do not seem to be an appropriate method to quantify the impact of saltmarsh

processes on coastal water productivity. Perhaps future research should concentrate on comparing nutrient concentrations, species composition and productivity of coastal water adjacent to saltmarshes with that remote from saltmarsh influence, in order to ascertain the importance of saltmarshes to coastal productivity.

Acknowledgements

The research was conducted at the University of East Anglia with the support of NERC funding

References

Axelrad, D.M. (1974) Nutrient flux through the salt marsh ecosystem. Ph.D Thesis, College of William and Mary, Virginia, USA.

Azni, S., Abd. Aziz, S. & Nedwell, D.B. (1986) The nitrogen cycle of an East Coast, U.K. saltmarsh: II. Nitrogen fixation, nitrification, denitrification, tidal exchange. *Estuarine, Coastal and Shelf Science* **22**, 689–704.

Boon, J. (III) (1975) Tidal discharge asymmetry in a saltmarsh drainage system. *Limnology and Oceanography* **20**, 71–80.

Boorman, L.A., Hazelden, J., Loveland, P.J., Wells, J.G. & Levasseur, J.E. (1994) Comparative relationships between primary productivity and organic and nutrient fluxes in four salt marshes. In *Global Wetlands; Old World and New*, Mitch W.J. (ed.), Elsevier, Amsterdam.

Burt, T.N. & Stevenson, J.R. (1986) Monitoring cohesive sediment transport in estuaries. In *International Conference on Measuring Techniques of Hydraulic Phenomena in Offshore, Coastal and Inland Waters, London, England 9–11 April 1986.*

Carpenter, K.E. (1993) Nutrient fluvial and groundwater flux between a North Norfolk saltmarsh and the North Sea. Ph.D Thesis, University of East Anglia, Norwich, UK.

Chambers, R.M. (1992) A fluctuating water level chamber for biogeochemical experiments in tidal marshes. *Estuaries* **15** (1), 53–8.

Dame, R.F., Spurrier, J.D., Williams, T.M., Kjerfve, B., Zingmark, R.G., Wolaver, T.G., Chrzanowski, T.H., McKellar, H.N. and Vernberg, F.J. (1991) Annual material processing by a salt marsh-estuarine basin in South Carolina, USA. *Marine Ecology Progress Series* **72**, 153–66.

Dankers, N., Binsbergen, M., Zegers, K., Laane, R. & van der Loeff, M.B. (1984) Transportation of water, particulate and dissolved organic and inorganic matter between a salt marsh and the Ems–Dollard Estuary, The Netherlands. *Estuarine, Coastal and Shelf Science* **19**, 143–65.

Day, J.W., Jr, Smith, W.G. & Wagner, P.R. (1973) Community structure and carbon budget of a salt marsh and shallow bay estuarine system in Louisiana. *Louisiana State University, Center of Wetland Resources. Publ. LSU-SG-72-04.* 74 pp.

Kjerfve, B. & Proehl, J.A. (1979) Velocity variability in a cross-section of a well mixed estuary. *Journal of Marine Research* **37**, 409–18.

Lee, V. (1979) Net nutrient flux between the emergent marsh and the tidal waters. MSc Thesis, University of Rhode Island, USA, 67 pp.

Marshall, D.E. (1970) Characteristics of Spartina marsh receiving municipal sewage waste. MSc thesis, University of Carolina, USA.

Nixon, S.W. (1980) Between coastal marshes and coastal water – a review of twenty years of speculation and research on the role of saltmarshes in estuarine productivity and water chemistry. In *Estuarine and Wetland Processes*, Hamilton, P. & Macdonald, K.B. (eds.), Plenum.

Odum, E.P. & de la Cruz, A.A. (1967) Particulate organic detritus in a Georgia salt marsh-estuarine ecosystem. In *Estuaries*, Lauff, G.H. (ed.), American Association for the Advancement of Science, Publication No. 83.

Patrick, W.H., Jr. & Delaune, R.O. (1976) Nitrogen and phosphorus utilization by *Spartina alterniflora* in a saltmarsh in Barataria Bay, Louisiana. *Estuarine, Coastal and Marine Science* **4**, 59–64.

Reed, D.J. (1987) Temporal sampling and discharge asymmetry in salt marsh creeks. *Estuarine, Coastal and Shelf Science* **25**.

Roman, C.T. (1984) Estimating water volume discharge through salt-marsh tidal channels: an aspect of material exchange. *Estuaries* **7**(3), 259–64.

Sargent, D.M. (1981) The development of a viable method of stream flow measurement using the integrating float technique. *Proc. Inst. Civ. Engrs.* **71** (part 2), 1–15.

Sargent, D.M. (1982a) The rising air float technique for the measurement of stream discharge. *Advances in Hydrometry* (proceedings of the Exeter Symposium, July 1982). IAHS Publ. no. 134.

Sargent, D.M. (1982b) The accuracy of streamflow measurement using the rising air float technique. *Proc. Inst. Civ. Engrs.* **73** (part 2), 179–82.

Scudlark, J.R. & Church, T.M. (1989) The sedimentary flux of nutrients at a Delaware saltmarsh site – a geochemical prospective. *Biogeochemistry* **7**(1), 55–75.

Teal, J.M. (1962) Energy flow in the saltmarsh ecosystem of Georgia. *Ecology* **43**, 614–24.

Tyler, G. (1967) On the effect of phosphorus and nitrogen, supplied to Baltic shore meadow vegetation. *Botanisk Notiser*. **120**, 433–47.

Valiela, I.B. & Teal, J.M. (1976) Primary production and dynamics of experimentally enriched saltmarsh vegetation: below ground biomass. *Limnology and Oceanography* **21**, 245–52.

Valiela I., Teal, J.M., Volkmann, S., Shafer, D. & Carpenter, E.D. (1978) Nutrient and particulate fluxes in a salt marsh ecosystem: tidal exchanges and inputs by precipitation and groundwater. *Limnology and Oceanography* **23**(4), 798–812.

Whiting, G.J., McKellar, H.N., Kjerfve, B. & Spurrier, J.D. (1985) Sampling and computational design of nutrient flux from a southeastern US saltmarsh. *Estuarine, Coastal and Shelf Science* **21**, 273–86.

Whiting, G.J., McKellar, H.N., Spurrier, J.D. & Wolaver, T.G. (1989) Nitrogen exchange between a portion of vegetated salt marsh and the adjoining creek. *Limnology and Oceanography* **34**(2), 463–73.

Wolaver, T.G., Zieman, J.C., Wetzel, R. & Webb, K.I. (1983) Tidal exchange of nitrogen and phosphorus between a mesohaline vegetated marsh and the surrounding estuary in the Lower Chesapeake Bay. *Estuarine, Coastal and Shelf Science* **16**, 321–32.

Wolaver, T.G. & Spurrier, J.D. (1988) The exchange of phosphorus between a Euhaline vegetated marsh and the adjacent tidal creek. *Estuarine, Coastal and Shelf Science* **26**, 203–14.

5

Nutrient recycling in intertidal sediments

S.J. Malcolm and D.B. Sivyer

5.1 Introduction

Intertidal sediments occupy a unique position between land and sea giving them an importance which is only now becoming apparent. Most studies have looked in detail at the nature of the biogeochemical processes (e.g. Klump & Martens, 1989) but have not had the impetus to place the processes into a quantitative context. The priority and scale of study has changed due to an awareness that intertidal areas are the front line in dealing with both the effects of climate change and the movement of materials from land to sea (Nowicki, 1994). There are now global, regional and national research programmes examining interactions at the land/sea boundary (e.g. LOICZ (Pernetta & Milliman, 1995), NECOP (Ortner & Dagg, 1995), JoNuS (Anon., 1994), LOIS (Anon., 1992)).

Concern about the impact of nutrients on the marine environment is providing a sharp focus for some of these studies (e.g. Nowicki & Oviatt, 1990; Anon., 1994). Intertidal sediments are a part of the sediment system that has long helped to maintain the productivity of the coastal zone via the storage and recycling of nutrients which were imported from offshore waters (Nixon, 1986; Nixon, 1992). Human population increases, coupled with aquatic sewerage systems and the development of intensive agriculture, have resulted in a significant change in the quantities of nutrients supplied from land sources. Now the main source of nutrients is from the land and not from the sea (Nixon, Hunt & Nowicki, 1986). What impact has this change had on the buffering role of intertidal sediments?

Intertidal areas are difficult to study from a practical point of view and offer conceptual challenges to developing the quantitative budgets required to address questions about nutrient transport between land and sea. The methods employed must take account of the key attributes of the intertidal environment which include variability, patchiness scales and dynamics. The biogeochemical processes that influence the behaviour of nutrients are reasonably well established but what controls them and how they interact with the physical environment is not well known.

The purpose of this paper is to outline the physical environmental factors involved in the functioning of the intertidal system and briefly review the impact on the biogeochemical processing of nitrogen, phosphorus and silicon. Consideration of saltmarshes is excluded. Discussion is focused on the often extensive areas of non-vegetated sand and mud flats that are uncovered by the tides on a daily basis and for which there is a limited amount of published information.

5.2 Physical processes impacting intertidal areas

The schematic shown in Figure 5.1 shows the major [bio]physical disturbances that impact on intertidal sediments. These are tides, waves, storms, run-off events and sediment movements which result from the above, bioturbation, bioirrigation and anthropogenic factors such as fishing and dredging. The disturbances occur over a wide range of time scales and influence processes over different space scales (Figure 5.2) but the net effect is to produce an environment that, although broadly structured, is patchy and dynamic.

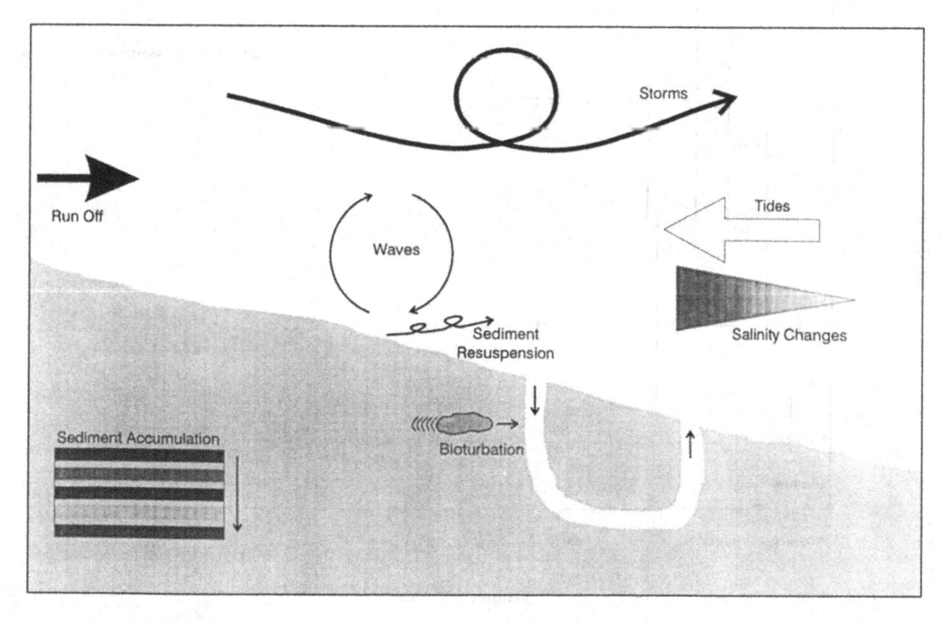

Figure 5.1 Schematic showing the [bio]physical impacts on intertidal sediments.

5.2.1 Tides

Intertidal sediments are subject to diurnal drying and flooding which leads to major changes in the local transport of material to and from the sediments. Clearly when the tide has ebbed there is no overlying water present and there can be no dissolved output from the sediment. This may result in increasing concentrations of dissolved substances that have a source within the sediments, such as ammonium. This pool of high concentration may then be released when the water returns to flood the sediment. In contrast, a substance whose source is in the water and is consumed in the sediment may become depleted when the water is not present. Examples of these processes are presented below and show that sediments are both sinks and sources of different nutrients and that the behaviour changes with the tide.

During the tidal cycle the hydraulic regime in the sediments changes and advective flow along pressure gradients may occur. Muddy intertidal sediments have a high porosity but have a low permeability limiting advective transport. Sandy sediments have a permeability great enough to allow drainage of the sediments at low water and subsequent saturation as

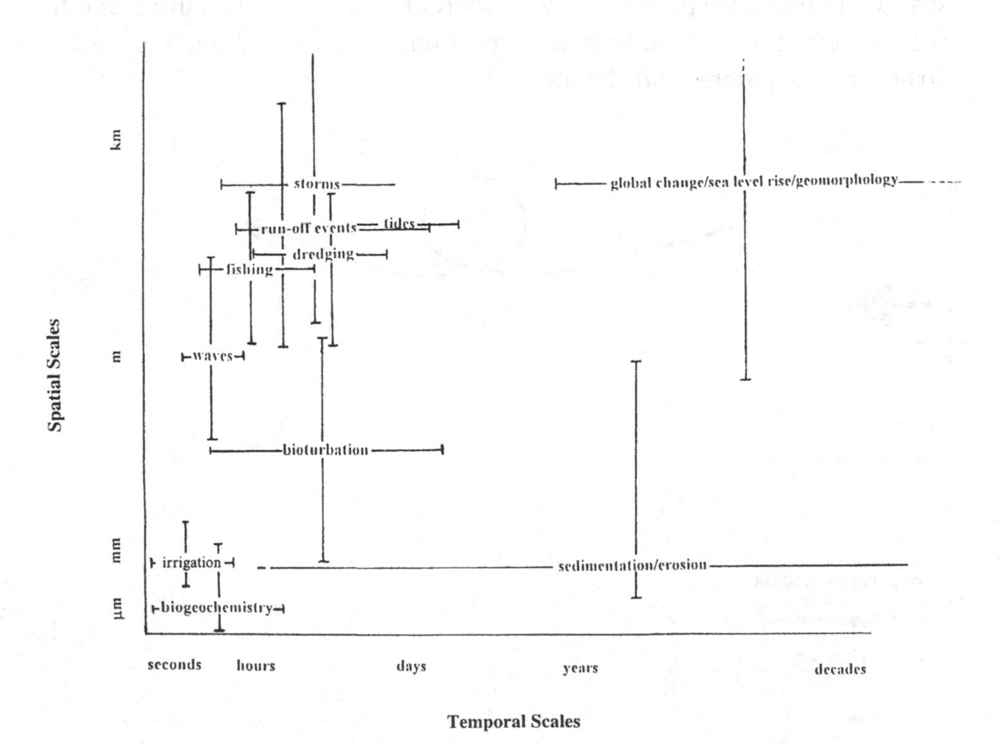

Figure 5.2 The spatial and temporaral scales of disturbance of the intertidal environment.

the tide advances. The impact of advection and resultant changes, such as increased penetration of oxygen into the sediments, has yet to be fully addressed but will have a major impact on redox sensitive processes such as coupled nitrification and denitrification (Rysgaard *et al.*, 1994).

Superimposed on the diurnal cycles, the tides exhibit a spring/neap cycle of changing tidal range which has two effects. Firstly, the area of sediment exposed and the length of exposure changes as the tide range alters and, secondly, the velocity of the ebb and flood currents changes. The sediment source/sink term is area related and the current velocity will have an impact on disturbance and transport of sediment and on flushing characteristics (Vorosmarty & Loder, 1994). Therefore, the export and import of material from intertidal areas is influenced by spring/neap cycles.

5.2.2 Storms/episodic events/waves

The impact of these infrequent events is a significant unknown. Tidal cycle changes are relatively easy to measure but the influence of events such as river spate (see below), storms and wind-induced wave action in shallow areas is more difficult to study. For safety reasons, it is not desirable to work in estuaries or on intertidal areas during storms and developments in this area will depend on autonomous measurement devices. Reliable instrumentation has only recently become available (e.g. Green *et al.*, 1992) and it will be some years before our understanding of the overall impact of episodic events is improved. However, a storm that disturbs much of the upper 10 cm of sediments over a given intertidal area is likely to release much of the associated dissolved nutrients which would be diluted into the water and flushed away. The loss of nutrient could be very large and would be different for each nutrient. Nitrogen and silicon would tend to be enriched in the water while phosphorus would remain attached to the sediment particles and re-enter the sediment system after the disturbance. Thus the sediments may accumulate phosphorus relative to the other nutrients and the concentration will be limited by the pseudo-equilibrium adsorption/desorption process which buffers phosphorus concentrations in marine and estuarine waters.

5.2.3 Run-off events

Major run-off events are a feature of most intertidal areas associated with estuaries. The high flows may limit nutrient processing due to decreased water residence times; however, if concentration increases with flow, nutrient processing may increase depending only on the biokinetics of the process. The concentration of nitrate in river water tends to increase with increasing water flow (Figure 5.3) as soluble nitrate is washed from soils by run-off (Rendell *et al.*, in press). Denitrification rate and extent are maximal during the winter when flows and resulting concentrations are high and may similarly be high during single high flow events at any time of year (Joye & Paerl, 1993).

5.2.4 Sediment transport

Intertidal sediments are generally dynamic. Although intertidal areas are normally accumulating sediment, the sediments are subject to time varying patterns of local erosion and deposition. The stability of these areas will determine the potential for nutrient storage. However, as the time scales of nutrient regeneration and recycling are generally fast, it may be possible to ignore sedimentation in the short term. However, leakage occurs from the surface sediment recycling into the underlying sediments. A relatively small percentage of the nitrogen and silicon (Brown, 1994) is incorporated into the sediments while significant amounts of phosphorus will remain stored (Fox, 1989; de Jonge Engelkes & Bekker, 1993; Prastka & Jickells, 1995). The extent of the sediment sink is an important factor in

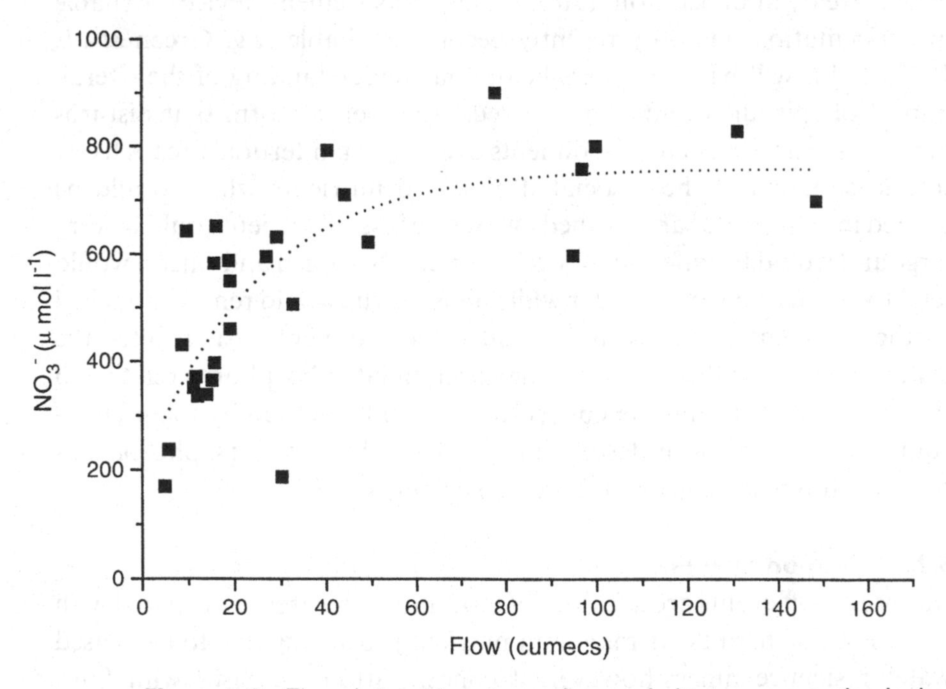

Figure 5.3 The relationship between flow and nitrate concentration in the river Great Ouse (UK) (JoNuS unpublished data). This relationship is typical of many European rivers draining agricultural catchments.

predicting the future impact of nutrients as measures are taken to control discharges to sea.

5.2.5 Bioirrigation and bioturbation

The macro-fauna that live in the sediments are one of the main ways of disturbing and restructuring intertidal sediments (Asmus, 1986). Burrowing animals will move through the sediment, causing homogenisation to a characteristic depth. Open burrow structures not only allow the movement of sediment from one depth to another but also substantially increase the sediment/water interface. The effect of this is greater oxygen penetration into the sediments and potentially greater export of regenerated nutrients (Clavero, Niell & Fernandez, 1994; Marinelli, 1994). Bioturbation has a significant impact on the rate of denitrification (Pelegri, Nielsen & Blackburn, 1994).

5.2.6 Human disturbance – fishing, dredging and reclamation

Fishing is thought (North Sea Task Force, 1993) to cause similar disturbance to that due to the passage of burrowing organisms but on a greater scale and at a greater rate, while dredging and subsequent disposal of the dredge spoil can be considered in the same category as a major storm event with substantial relocation of particulate material. On a much longer time scale, reclamation and coastal defence engineering which reduces the intertidal area will have a dramatic effect but this will not be considered here other than to say that, although there are few examples of this impact in the literature, it is an area of major importance as awareness increases of the impact of engineering management on estuaries and coastal systems.

5.3 Nutrient processing

Nutrients are affected by a variety of processes in intertidal sediments (Klump & Martens, 1989). Nitrogen is particularly interesting as it exists in different forms, both organic and inorganic, and may be converted from one form to another under different conditions. Though subject to continuing debate, nitrogen is also considered to be the nutrient limiting primary production in the sea (Gilbert, 1988; Owens, 1993). This has led to a substantial literature describing aspects of the nitrogen cycle (Blackburn & Sorensen (1988) and references therein). The behaviour of phosphorus is dominated by its propensity to associate with particles and iron oxides in the estuarine and coastal environment. The fate of phosphorus is bound up with the movement of sediments but the details of phosphorus cycling at the local and global level are only now emerging (Berner *et al.*, 1994). Silicon geochemistry is considered to be relatively simple; silicon is removed from water by diatoms and temporarily stored in sediments as biogenic opal (Berner, 1980; Gehlen & van Raaphorst, 1993). The dynamics of silicon behaviour have not been well studied and the growth of diatoms in mats (de Jonge, 1980) at the sediment surface are likely to be an important component of the silicon story.

5.3.1 Nitrogen

Nitrogen can be present in a large variety of forms in aquatic systems. The main species of interest are nitrate, nitrite and ammonium together with organic nitrogen, in both dissolved and particulate form. There is a large pool of di-nitrogen gas in the atmosphere and dissolved in surface waters but it is considered that nitrogen fixation is not important in coastal systems (Seitzinger, 1988). Inputs of nitrogen to intertidal sediments come from the land, in solution via rivers or groundwater, and in the form of particulate material following uptake by phytoplankton. Regeneration of the material that escapes the pelagic system is thought to be the principal source for the nutrients that support plankton growth through the

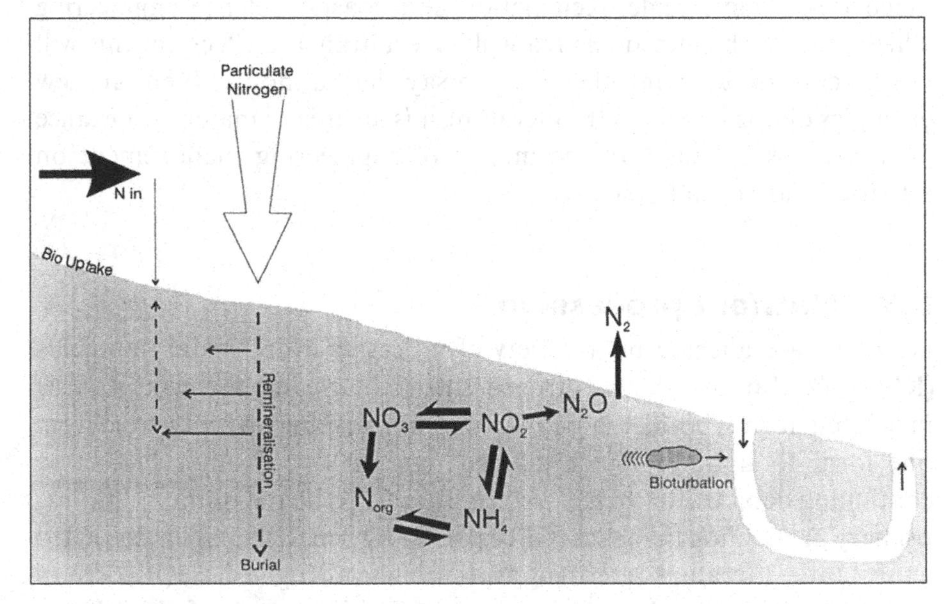

Figure 5.4 Schematic showing the key nitrogen cycle processes in intertidal sediments.

summer when direct input from land is low. Figure 5.4 shows a schematic for the nitrogen cycle in intertidal sediments. What makes these sediments different is not the nature of the individual processes but the relative importance of each and the way they are controlled by the unique physical environment.

Nitrogen fixation

Nitrogen fixation converts di-nitrogen gas to fixed nitrogen and is the main return path of nitrogen from the atmosphere to the biosphere. In the developed parts of the world anthropogenic nitrogen fixation (Kinzig & Socolow, 1994) dominates the nitrogen cycle. There have been many studies of the process in the coastal marine environment. However, due to the ready availability of ammonium and nitrate in intertidal areas, the nitrogen requirements of plants are not provided by nitrogen fixation (Capone, 1988). Nitrogen fixation will be important where exogenous inputs of nitrogen are low or when the pool of readily available nitrogen is depleted by biological production.

Nitrification and denitrification

These processes and their links have been the focus for much scientific effort (Koike & Sorensen, 1988), not only because they are key nitrogen transformation processes but also because they provide the most important route for removal of nitrogen from ecosystems. Nitrification is the oxidation of ammonium to nitrate, where the ammonium is produced from organic nitrogen. These are key steps in the regeneration of nitrogen. Fluxes of both ammonium, which is the preferred source of nitrogen for phytoplankton, and nitrate from sediments to the water have been measured. Under appropriate conditions, and there is some debate about those conditions, nitrate can be converted to gaseous products, principally di-nitrogen (Seitzinger, 1980). Coupled nitrification/denitrification has been considered to be the dominant source of gaseous nitrogen products in sediments (Lloyd, 1993) but recent work (e.g. Rysgaard *et al.*, 1994) suggests that under high nitrate concentration denitrification direct from nitrate is the major pathway. This is illustrated in Figure 5.5, using unpublished data from the JoNuS programme, which shows the progressive loss of nitrate from sediment interstitial water during exposure of intertidal sediments. This is particularly significant when considering the flow of nitrogen from land to sea in systems that are heavily impacted by nitrate run-off and, serendipitously, provides a mechanism for amelioration of the impact in coastal seas which may be at risk of problems resulting from eutrophication.

Nitrous oxide

Nitrous oxide is a product of both nitrification and denitrification and is of environmental concern as it is an important greenhouse gas. Law, Rees & Owens (1992) demonstrate the potential importance of estuaries as a source of N_2O to the atmosphere. The yield of nitrous oxide during denitrification in marine and estuarine systems is of the order of a few per cent of the total denitrification flux, much less than the relative yield from denitrification in fresh water and terrestrial systems (20–50%). Thus in considering effective management policies, and assuming a desire to avoid adding more N_2O to the atmosphere, brackish water and intertidal environments may offer a better nitrogen removal system rather than encouraging extensive use of fresh water wetlands (e.g. Fleischer, Stibe & Leonardson, 1991).

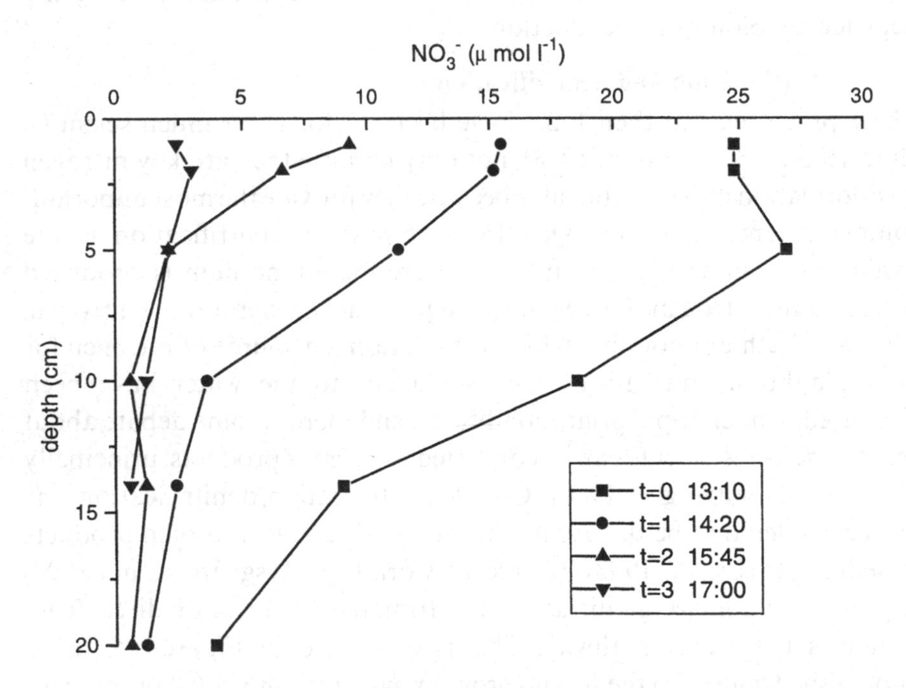

Figure 5.5 Time series of interstitial water profiles at an intertidal site in the Wash (UK), near to the major riverine discharge, after the tide has ebbed. The inventory of nitrate in the interstitial water decreases with time due to rapid denitrification. The interstitial water profiles are collected using an *in-situ* 'sipping' system consisting of porous probes inserted at fixed depths in the sediment from which water samples are removed by applying a vacuum. Samples are preserved using mercuric chloride and returned to the laboratory for colorimetric analysis.

5.3.2 Phosphorus

Compared with nitrogen, the behaviour of phosphorus in intertidal sediments is less well known. Phosphorus associates with solid surfaces, particularly the surface of iron oxides in sediments (Prastka & Malcolm, 1994), developing a pseudo-equilibrium which acts as a buffer mechanism in estuaries (Van Beusekom & de Jonge, 1994). Phosphorus is released under anoxic conditions by the degradation of organic material and accumulates in the interstitial water. In intertidal sediments with a low organic carbon and phosphorus content the behaviour of phosphorus is dominated by the association with iron oxides (Prastka & Jickells, 1995) and is also released to the interstitial water under anoxia. However, remobilisation from the sediments is tempered by the fact that under oxic conditions, which occur at the sediment/water interface, phosphorus binds strongly to particles. Thus, to a first approximation, intertidal sediments act as a significant sink for phosphorus. Disturbance of the sedimentary regime is the only mechanism that will allow leakage of phosphorus from the system.

5.3.3 Silicon

Input of silicon to intertidal sediments comes from two sources, the sedimentation of the tests of pelagic diatoms and *in situ* uptake by benthic diatoms from the water. Both processes result in the input of particulate biogenic silica to the sediment which corrodes in alkaline sea water (Barker *et al.*, 1994). Silicon accumulates in the interstitial water which results in a flux of silicon out of the sediments. The flux may have a significant seasonal variation due to the growth of benthic diatoms during the spring/summer period and due to bioirrigation (Marinelli, 1994).

Diatoms have been used to look at the history of changing productivity, and hence nutrient inputs, in lake systems. The use of diatoms as palaeoenvironmental indicators in coastal environments is limited because of the preservation problem (Barker *et al.*, 1994) and the paucity of areas which are continuously accreting sediment. Brown (1994) studied a number of sediment cores from the Wash for preservation of diatom material. The biogenic silica content of the sediments decreases with depth (Figure 5.6) which may indicate progressive corrosion with depth and/or increased input of silica to the sediments with time. Gross changes in the type of diatom present in the sediment may support the latter conclusion but the work is of a preliminary nature. Further examination of siliceous palaeoenvironmental indicators in intertidal sediments will be useful.

5.4 Benthic mats

An important feature of intertidal sediments is the presence of a mat of diatoms and associated bacteria in the surface sediment. Benthic production in clear shallow water can be important, as benthic productivity can equal or exceed the productivity of the phytoplankton. This will clearly have a major impact on nutrients by direct uptake, adding to the pool of organic material available for regeneration as well as modifying the transport of nutrients between sediment and water (Hopner & Wonneberger, 1985). The presence of a mat also helps to stabilise the sediment surface making it less prone to physical disturbance. Mats of benthic microphytes produce and consume oxygen in the surface sediments, enhancing the penetration of oxygen into the sediments during the photo period. The alteration of the oxygen regime has a significant impact on both the pathway and extent of denitrification (Sundback *et al.*, 1991; Risgaard-Petersen *et al.*, 1994). The whole balance of this system could therefore change dramatically between summer when production is occurring and winter when the organic material produced in the summer is being degraded. This could enhance the amount of denitrification from nitrate in winter run-off by bringing the redoxcline nearer to the sediment surface and enhance

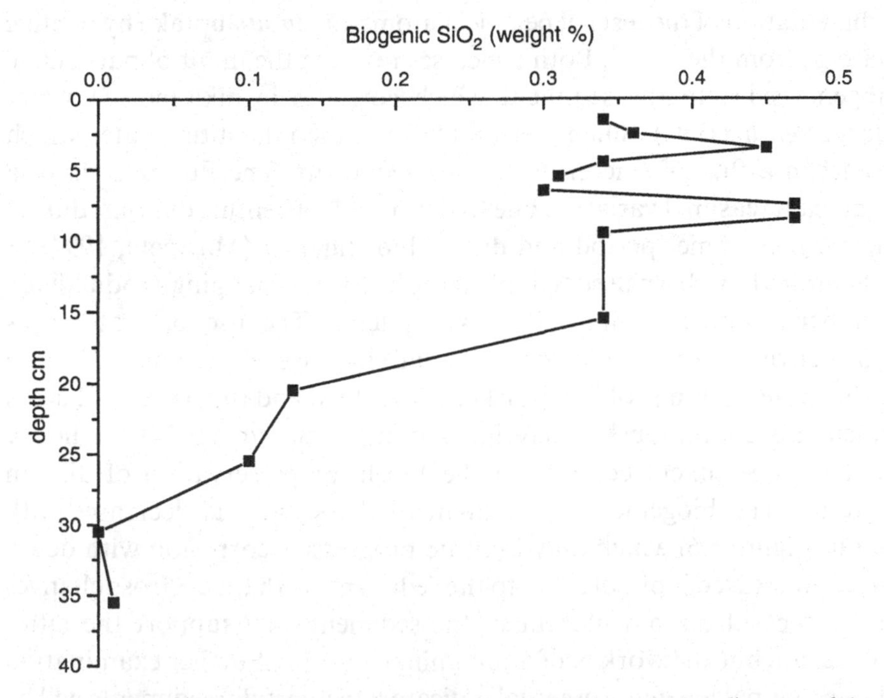

Figure 5.6 Distribution of biogenic silica in intertidal sediments of the Wash. Biogenic silica was determined by a wet chemical method.

coupled nitrification/denitrification in the summer by increasing the oxygen supply.

5.4.1 Morphological control of the impact of intertidal areas

If the processes occurring in intertidal sediments have some unique qualities and result in sometimes unexpected impacts on nutrients, there is one feature which may dominate the overall importance of intertidal areas. This is the morphology of the system. For example, the intertidal sediments of the Wash embayment located at the distal end of the canalised Gt Ouse Estuary (Figure 5.7*a*) remove a minimum of 30% of the incident nitrate by denitrification. This may be contrasted with the Humber Estuary where the intertidal areas are located marginally (Figure 5.7*b*) and the overall loss is no more than 10% (unpublished JoNuS data). Although other factors have a role, it is the location of the intertidal areas with respect to the flow of water that is critical in determining the impact of intertidal sediments on the transport of nutrients from land to sea.

5.5 Concluding remarks

Intertidal sediments occupy a unique position to exert a controlling influence on the transfer and recycling of nutrients between land and sea.

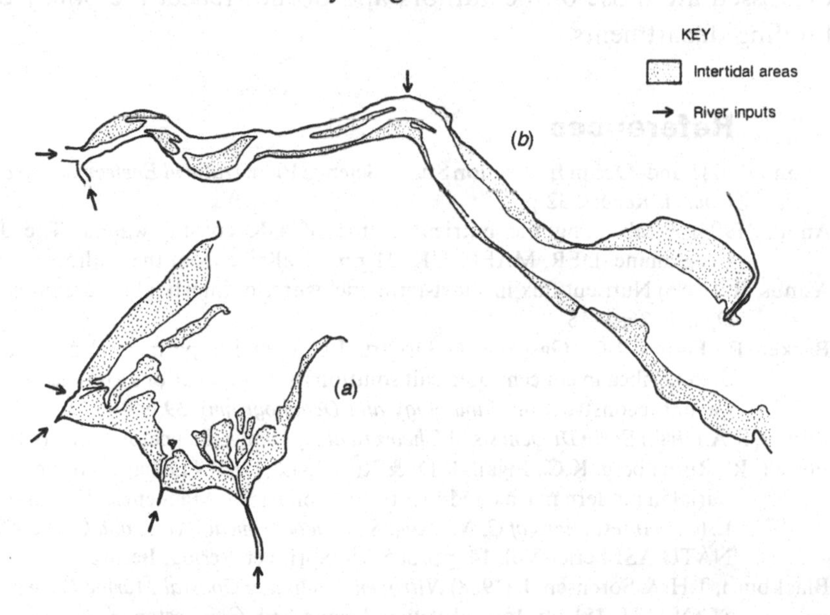

Figure 5.7 The distribution of intertidal sediments in (*a*) the Wash and (*b*) the Humber Estuary in relation to the main fresh water inputs. The intertidal in the Wash is at the distal end of canalised estuaries whereas the intertidal in the Humber tends to be marginal to the flow of water. Diagram not to scale.

Prior to anthropogenic change of global nutrient cycles it is likely that intertidal sediments were importers of nutrients from the sea. The productivity of coastal waters was maintained by storing and regenerating nutrients on a seasonal basis allowing the development of significant fisheries. The main inputs of nutrients are now land derived due to the impact of human activities. Intertidal areas may now be acting as a buffer which, while maintaining the productivity of coastal waters, acts to ameliorate inputs of nutrients to the coastal seas. The capacity of the buffer system is not known but must be considered when management measures designed to alleviate problems caused by nutrient input are being formulated.

Acknowledgements

The authors would like to thank all of the members of the JoNuS Project Team for interesting discussions about nutrient transformations in estuaries and coastal environments. JoNuS (Joint Nutrient Study) seeks to quantify the transport of nutrients from land to sea and is funded by the Ministry of Agriculture, Fisheries and Food, the Department of the Environment and the National Rivers Authority (all UK). The views expressed are those of the authors and do not reflect the policy of the funding departments.

References

Anon. (1992) Land–Ocean Interaction Study, Science Plan. *Natural Environment Research Council Report*, 32 pp.

Anon. (1994) Establishing the nutrient status of UK coastal waters: The JoNuS Programme. DFR, MAFF, UK. 31 pp. (available from the authors).

Asmus, R. (1986) Nutrient flux in short-term enclosures of intertidal sand communities. *Ophelia* **26**, 1–18.

Barker, P., Fontes, J-C., Gasse, F. & Druart, J-C. (1994) Experimental dissolution of diatom silica in concentrated salt solutions and implications for palaeoenvironmental reconstruction. *Limnology and Oceanography* **39**, 99–110.

Berner, R.A. (1980) *Early Diagenesis: A Theoretical Approach.* Princeton University Press.

Berner, R., Ruttenberg, K.C., Ingall, E.D. & Rao, J.-L. (1984) The nature of phosphorus burial in modern marine sediments. In Wollast, R., Mackenzie, F.T. and Chou, L. (eds.) *Interactions of C, N, P and S Biogeochemical Cycles and Global Change.* NATO ASI Series Vol. 14, pp. 365–78. Springer-Verlag, Berlin.

Blackburn, T.H. & Sorensen, J. (1988) *Nitrogen Cycling in Coastal Marine Environments.* SCOPE 33, 451 pp. John Wiley and Sons Ltd, Chichester.

Brown L.J. (1994) The geochemistry of biogenic silica in sediments. MSc. Thesis, University of Leeds, Leeds, UK.

Capone, D.G. (1988) Benthic nitrogen fixation. In Blackburn, T.H. and Sorensen, J. (eds.)

Nitrogen Cycling in Coastal Marine Environments. SCOPE 33, pp. 85–123. John Wiley and Sons Ltd, Chichester.

Clavero, V., Niell, F.X. & Fernandez, J.A. (1994) A laboratory study to quantify the influence of *Nereis diversicolor* O.F. Muller in the exchange of phosphate between sediment and water. *Journal of Experimental Marine Biology and Ecology* **176**, 257–67.

Fleischer, S., Stibe, L. & Leonardson, L. (1991) Restoration of wetlands as a means of reducing nitrogen transport to coastal waters. *Ambio* **20**, 271–2.

Fox, L.E. (1989) A model for inorganic control of phosphate concentrations in river waters. *Geochim. Cosmochim. Acta* **53**, 417–28.

Gehlen, M. & Raaphorst, W. van (1993) Early diagenesis of silica in sandy North Sea sediments: quantification of the solid phase. *Mar. Chem.* **42**, 71–83.

Gilbert, P.M. (1988) Primary productivity and pelagic nitrogen cycling. In Blackburn, T.H. and Sorensen, J. (eds.) *Nitrogen Cycling in Coastal Marine Environments.* SCOPE 33, pp. 1–31. John Wiley and Sons Ltd, Chichester.

Green, M.O., Pearson, N.D., Thomas, M.R., Rees, C.D., Rees, J.R. & Owen, T.R.E. (1992) Design of a data logger and instrument platform for seabed sediment transport research. *Continental Shelf Research* **12**, 543–62.

Hopner, T. & Wonneberger, K. (1985) Examination of the connection between the patchiness of benthic nutrient efflux and epiphytobenthos patchiness on intertidal flats. *Netherlands Journal of Sea Research* **19**, 277–85.

Jonge, V.N. de (1980) Fluctuations in the organic carbon to chlorophyll ratios for estuarine benthic diatom populations. *Mar. Ecol. Prog. Ser.* **2**, 345–53.

Jonge, V.N. de, Engelkes, M.M. & Bakker, J.F. (1993) Bio-availability of phosphorus in sediments of the western Dutch Wadden Sea. *Hydrobiologia* **253**, 151–63.

Joye, S.B. & Paerl, H.W. (1993) Contemporaneous nitrogen fixation and denitrification in intertidal microbial mats: rapid response to runoff events. *Marine Ecology Progress Series* **94**, 267–74.

Kinzig, A.P. & Socolow, R.H. (1994) Human impacts on the nitrogen cycle. *Physics Today* **46**, 24–31.

Klump, J.V. & Martens, C.S. (1989) The seasonality of nutrient regeneration in an organic rich coastal sediment: kinetic modeling of changing pore-water nutrient and sulfate distributions. *Limnology and Oceanography* **34**, 559–77.

Koike, I. & Sorensen, J. (1988) Nitrate reduction and denitrification in marine sediments. In Blackburn, T.H. and Sorensen, J. (eds.) *Nitrogen Cycling in Coastal Marine Environments.* SCOPE 33, pp. 251–73. John Wiley and Sons Ltd, Chichester.

Law, C.S., Rees, A.P. & Owens, N.J.P. (1992) Nitrous oxide: estuarine sources and atmospheric flux. *Estuarine, Coastal and Shelf Science* **35**, 301–14.

Lloyd, D. (1993) Aerobic denitrification in soils and sediments: from fallacies to facts. *TREE* **8**, 352–6.

Marinelli, R.L. (1994) Effects of burrow ventilation on activities of a terebellid polychaete and silicate removal from silicate pore waters. *Limnology and Oceanography* **39**, 303–17.

Nixon, S.W. (1986) Nutrient dynamics and the productivity of coastal waters. In Halwagy, R., Clayton, D. and Behbehani, M. (eds.) *Marine Environment and Pollution*, pp. 97–115. The Alden Press, Oxford.

Nixon, S.W. (1992) Quantifying the relationship between nitrogen input and the productivity of marine ecosystems. *Proceed. Adv. Mar. Tech. Conf., Tokyo* **5**, 57–83.

Nixon, S.W., Hunt, C.D. & Nowicki, B.L. (1986) The retention of nutrients (C, N, P), heavy metals (Mn, Cd, Pb, Cu) and petroleum hydrocarbons in Narragansett Bay. In

Lasserre, P. and Martin, J.-M. (eds.) *Biogeochemical Processes at the Land/Sea Boundary*, pp. 99–122. Elsevier, London.
North Sea Task Force (1993) *North Sea Quality Status Report*. Oslo and Paris Commissions, London. Olsen and Olsen, Fredensborg, Denmark, 132 pp.
Nowicki, B.L. (1994) The effect of temperature, oxygen, salinity and nutrient enrichment on estuarine denitrification rates measured with a modified nitrogen gas flux technique. *Estuarine, Coastal and Shelf Science* **38**, 137–56.
Nowicki, B.L. & Oviatt, C.A. (1990) Are estuaries traps for anthropogenic nutrients? Evidence from estuarine mesocosms. *Marine Ecology Progress Series* **66**, 131–46.
Ortner, P.B. & Dagg, M.J. (1995) Nutrient-enhanced coastal ocean productivity explored in the Gulf of Mexico. *EOS, Transact. Amer. Geophs. Union* **76**, 98–109.
Owens, N.J.P. (1993) Nitrate cycling in marine waters. In Burt, T.P., Heathwaite, A.L. and Trudgill, S.T. (eds.) *Nitrate: Processes, Patterns and Management*, pp. 169–209. John Wiley and Sons Ltd., Chichester.
Pelegri, S.P., Nielsen, L.P. & Blackburn, T.H. (1994) Denitrification in estuarine sediment stimulated by the irrigation activity of the amphipod *Corophium volutator*. *Marine Ecology Progress Series* **105**, 285–90.
Pernetta, J.C. & Milliman, J.D. (1995) Land Ocean Interactions in the Coastal Zone. Implementation Plan. *IGBP Global Change Report* **33**, 215 pp. International Council of Scientific Unions.
Prastka, K.E. & Jickells, T.D. (1995) Sediment/water exchange of phosphorus at two intertidal sites on the Great Ouse estuary, S.E. England. *Netherlands Journal of Aquatic Ecology* **29**, 245–55.
Prastka, K.E. & Malcolm, S.J. (1994) Particulate phosphorus in the Humber estuary. *Netherlands Journal of Aquatic Ecology* **28**, 397–403.
Rendell, A.R., Horrobin, T.M., Jickells, T.D., Edmunds, H.M., Brown, J. & Malcolm, S.J. (in press) Nutrient cycling in the Great Ouse Estuary and its impact on nutrient fluxes to The Wash, England. *Estuarine, Coastal and Shelf Science*.
Risgaard-Petersen, N., Rysgaard, S., Nielsen, L.P. & Revsbech, N.P. (1994) Diurnal variation of denitrification and nitrification in sediments colonized by benthic microphytes. *Limnology and Oceanography* **39**, 573–9.
Rysgaard, S., Risgaard-Petersen, N., Sloth, N.P., Jensen, K. & Nielsen, L.P. (1994) Oxygen regulation of nitrification and denitrification in sediments. *Limnology and Oceanography* **39**, 1643–52.
Seitzinger, S.P. (1980) Denitrification and N_2O production in nearshore marine sediments. *Geochim. Cosmochim. Acta.* **44**, 1853–60.
Seitzinger, S.P. (1988) Denitrification in freshwater and coastal marine ecosystems: ecological and geochemical significance. *Limnology and Oceanography* **33**, 702–24.
Sundback, K., Enoksson, V., Graneli, W. & Pettersson, K. (1991) Influence of sublittoral microphytobenthos on the oxygen and nutrient flux between sediment and water: a laboratory continuous flow study. *Marine Ecology Progress Series* **74**, 263–79.
Van Beusekom, J.E.E. & de Jonge, V.N. (1994) The role of suspended matter in the distribution of dissolved inorganic phosphate, iron and aluminium in the Ems estuary. *Netherlands Journal of Aquatic Ecology* **28**, 383–95.
Vorosmarty, C.J. & Loder, T.C. (1994) Spring–Neap tidal contrasts and nutrient dynamics in a marsh-dominated estuary. *Estuaries* **17**, 537–51.

6

○ ○ ○ ○ ○ ○ ○ ○ ○ ○ ○ ○ ○ ○ ○ ○ ○ ○ ○ ○

An overview of carbon and sulphur cycling in marine sediments

G. Ruddy

6.1 Introduction

In most marine-type sediments (including intertidal ones) organic carbon is the only reducing agent to enter a sediment column. The remainder of the sediment load arrives in its oxidised form, and, with the exception of straightforward compaction, early diagenesis (i.e. the process of change during burial) results directly or indirectly from the flow of electrons. The initial source of the electrons (organic matter) is sequentially oxidised in microbially mediated reactions, using a range of available oxidising agents, and results in some degree of vertical zonation in sediment chemistry (Richards *et al.*, 1965; Froelich *et al.*, 1979). This is because microbial communities outcompete each other for organic carbon (Stumm & Morgan, 1970; Claypool & Kaplan, 1974). However, the resolution of these sediment horizons is likely to be poor because:

(a) although the most important substrate shared by each community is organic carbon, it has been suggested that different communities metabolise different fractions of it, each with a first order rate constant (Berner, 1977). This would produce overlapping or coexistence of zones, unless the competition between communities was for another shared substrate (e.g. nitrate).

(b) the 'dead cat in the mud' effect (Coleman, 1985), where the substrates for microbial metabolism do not enter the system as disseminated and reactive particles to give a homogeneous reaction mixture, but rather as poorly sorted, highly concentrated sources of metabolites that may survive beyond zones where they would have been exhausted in a simple layered model. For solid substrates such as organic carbon, iron oxides and manganese oxyhydroxides, this may be an important control on their rate of supply to the reaction.

(c) the rate of supply of soluble oxidising agents (i.e. oxygen, nitrates and sulphate) and the removal of reaction products (e.g. H_2S, Fe^{2+}, Mn^{2+}, HCO_3^-) will depend not only on the rate of burial and diffusion (Lerman, 1979), but also on the modifying effects of bioturbation (e.g. Aller, 1980, 1984) and the regeneration of oxidising potential in oxidising conditions (Goldhaber & Kaplan, 1974; Jorgensen, 1977; Swider & Makin, 1989). Biomixing (e.g. Berner & Westrich, 1985; Hines & Jones, 1985; Sharma *et al.*, 1987) and tortuosity (the lengthening of diffusion paths in sediment pores relative to open solutions, e.g. Li & Gregory, 1974) may therefore have important consequences for the arrangement of the diagenetic zones.

(d) seasonality will cause the sinusoidal migration of zones upwards and downwards from an average or net level due to factors which vary seasonally such as bioturbation and temperature (Jorgensen, 1977; Abdollahi & Nedwell, 1979; Troelson & Jorgensen, 1982), and hence each reaction will move constantly through the diagenetic effects of other reactions. The resolution of boundaries may be lowered further by different rates of equilibrium attainment for the different components of each zone.

Where the sedimentation rate is low, or other flux controls such as the rate of diffusion are high, the observed zonation may depend very little on the burial fluxes of reactants, and the chemical system may be considered relatively open to exchange with its surroundings. The greater this degree of openness is, the greater the perturbation of a layered sequence of diagentic reactions is likely to be. This is likely to be particularly true of intertidal sediments and has important consequences for the interpretation of observations.

What follows is a discussion of some of the carbon and sulphur mechanisms and ideas contained within the diagenetic literature. It is intended to give the reader both a general background against which to interpret observations and also to emphasise the importance of micro-structure in understanding natural systems.

6.2 Carbon cycling

The rate of organic carbon remineralisation by microbial oxidation has been empirically shown by Middleburg (1989) to follow a first order decay curve. This relationship accounts for carbon diagenesis over about eight orders of magnitude in the carbon influx, covering very diverse environments.

The relationship can be stoichiometrically related to rates of electron acceptor reduction and has the advantage over previous, more complicated models, such as Berner's 'multi-G' model (Berner, 1978) in that no distinction between different types of organic carbon with different reactivities, is required. If the relationship is valid, it suggests that carbon diagenesis rates are determined by one common reaction. The suggestion is that this reaction is the bacterial fermentation of bio-polymers, producing metabolisable organic carbon for microbial communities using electron acceptors like oxygen and sulphate. This step produces amino acids, fatty acids, sugars, etc. (collectively known as monomers) from the proteins, lipids and carbohydrates which comprise the labile biopolymers (Welte, 1973). This transformation is similar to methanogenesis in that no external oxidising agent is involved (i.e. it is a form of chemical disproportionation).

The rates of monomer metabolism also appear to be first order with respect to the concentration of the monomer (rather than second order with respect to monomer and oxidising agent concentration), and therefore the concentration of the oxidising agent in question can be regarded as being in excess within its 'zone' (Devol *et al.*, 1984; Martens & Klump, 1984; Kuivila, Murray & Devol, 1989). This suggests that microbial communities do indeed compete for metabolisable organic carbon.

In Table 6.1 a series of idealised decomposition reactions are listed in order of their energy yields per mole of carbon, and therefore, broadly speaking, their diagenetic sequence. The relative importance of each reaction to the overall carbon budget depends on the relative supply of each reaction's electron acceptor and the relative supply of each reaction's metabolisable carbon. For carbon supply, the rate of sedimentation is

Table 6.1 Idealised decomposition reactions in order of their energy yields per mole of carbon from an idealised organic molecule (after Richards *et al.*, 1965)

Aerobic respiration: $(CH_2O)_x(NH_3)_y(H_3PO_4)_z + (x + 2y)O_2 \Rightarrow xCO_2 + (x + 1y)H_2O + yHNO_3 + zH_3PO_4$

Manganese reduction: $(CH_2O)_x(NH_3)_y(H_3PO_4)_z + 4xMnOOH + 7xCO_2 + xH_2O \Rightarrow 8xHCO_3^- + 4xMn^{2+} + yNH_3 + zH_3PO_4$

Nitrate reduction: $(CH_2O)_x(NH_3)_y(H_3PO_4)_z + 4xNO_3^- \Rightarrow xCO_2 + 3xH_2O + 4xHCO_3^- + 2xN_2 + 5yNH_3 + 5zH_3PO_4$

Iron reduction: $(CH_2O)_x(NH_3)_y(H_3PO_4)_z + 4xFe(OH)_3 + 7xCO_2 \Rightarrow 8xHCO_3^- + 3xH_2O + 4xFe^{2+} + yNH_3 + zH_3PO_4$

Sulphate reduction: $2(CH_2O)_x(NH_3)_y(H_3PO_4)_z + xSO_4^{2-} \Rightarrow 2xHCO_3^- + xH_2S + 2yNH_3 + 2xH_3PO_4$

Methane production: $2(CH_2O)_x(NH_3)_y(H_3PO_4)_z \Rightarrow xCO_2 + xCH_4 + 2yNH_3 + 2zH_3PO_4$

important, since a rapid accretion rate will tend to increase the flux of labile carbon to lower 'zones' in the sediment column and therefore enhance their contribution to overall remineralisation. Clearly the size of the influx will also be important in this respect, since a low flux will tend to be used up more aerobically than a bountiful one for a given accretion rate. Rates of supply of the various electron acceptors vary depending on:

(a) their rate of burial
(b) particulate nature
(c) molecular diffusion coefficients
(d) resupply by biomixing from above
(e) regeneration of their oxidising potential by migration into more oxidising environments.

Regeneration processes, in particular, are considered to be critical during early diagenesis, since they will usually account for the bulk of potential energy release (e.g. Ruddy, 1993). It is in this respect that micro-environments have an important role to play in sediment chemistry.

A mass balance model for carbon based on the observations of many workers in this field is shown in Figure 6.1. It is designed to give some indication of the types of profiles that might be expected in a typical sediment column from a mass balance point of view, rather than to

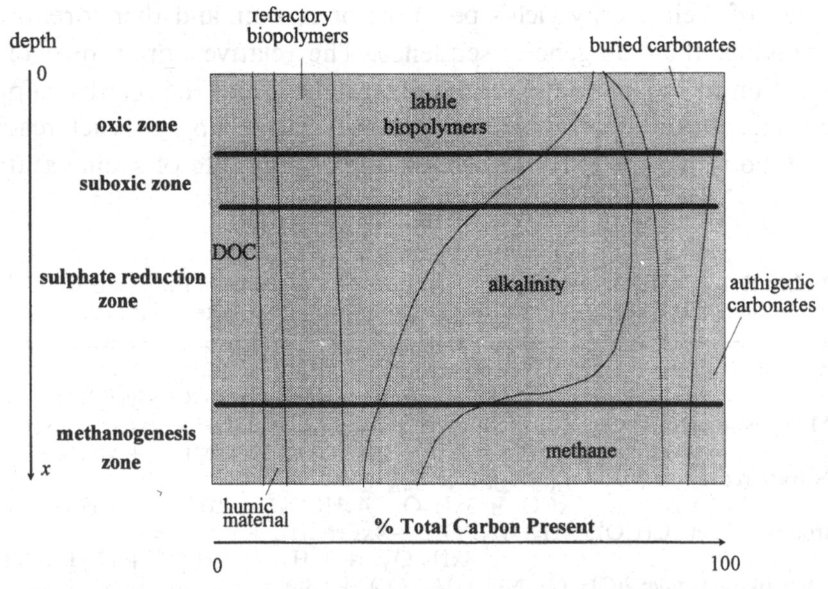

Figure 6.1 Notional carbon speciation in sediments. Profiles of carbon reservoirs observed in sediments, combined to give a qualitative picture of carbon cycling processes.

quantify each carbon reservoir. In the oxic layer, decomposition proceeds aerobically. Aller (1982) suggests that biomixing of labile organic carbon downwards will result in a 'convex up' profile in this zone, and also prevents the build up of alkalinity due to the enhanced upward diffusion of dissolved carbon dioxide. If the oxic layer is well established then much of the diffusive flux of carbon dioxide out of the sediment column will originate from this zone (Janke, Emerson & Murray, 1982; McNichol, Lee & Druffel, 1988).

Suboxic metabolism, which includes nitrate, manganese and iron reduction, does not generally seem to form a discrete layer due to the relatively low levels of nitrate available under natural conditions and the insoluble nature of manganese oxyhydroxides and iron oxides. These reactions appear to be restricted, where they are apparent at all, to micro-environments, but since, of the reduction reactions in Table 6.1, they are the only ones to generate alkaline conditions, they may be important for the precipitation and preservation of carbonate minerals.

Sulphate reduction in marine sediments is often the dominant form of carbon remineralisation, and contributes most of the alkalinity observed in the sediment pore-water. When concentrations of sulphate fall below 35 to 40 μM, carbon dioxide reduction by methanogens may begin, whereas sulphate levels below about 30 μM are required before the onset of acetate-type reduction reactions that dominate methanogenesis (Kuivila *et al.*, 1989). Carbon dioxide reduction is usually only evident in the sulphate reduction zone, where bicarbonate is produced, but it can contribute about 65% of the methane flux at its peak (Crill & Martens, 1986).

Under certain circumstances, the bulk of the methane flux leaving the sediment does so as gas bubbles, which can form established tubes through the sediment in summer months, provided the hydrostatic pressure is low enough (Martens & Klump, 1984). There have also been many observations of a secondary sulphate reduction rate peak at the sulphate reduction–methanogenesis interface (e.g. Crill & Martens, 1987). In this case the effect appeared to occur as summer blooms of short duration, in close association with hydrogen-oxidising communities in the sulphate reducing zone, and may have accounted for as much as 30% of the integrated yearly sulphate reduction where it was present.

At steady state, the carbon leaving the system as carbon dioxide and methane fluxes into the overlying water or air, will be equivalent to the amount of carbon remineralised in the column below, integrated over its depth. Dissolved organic carbon (DOC) exchange from pore-water to sea water is thought to have a minor effect on the mass balance (Martens &

Klump, 1984). Howes, Dacey & King (1984) have shown that saltmarsh diagenesis of organic carbon produced broadly similar carbon remineralisation and oxygen uptake rates (not because the carbon was aerobically remineralised, but because the oxygen uptake also integrated the oxidation of reduced products and therefore anaerobic remineralisation as well). This budget, however, incorporated some important seasonal fluctuations, where during the summer, carbon remineralisation rates were higher than oxidation rates and resulted in the build up of reduced end products. During the winter the trend was reversed with a corresponding increase in the system's oxidising potential.

The reverse situation is observed at Stiffkey marsh, North Norfolk, UK (Ruddy, 1993). Here, the steady state decomposition of organic carbon (Figure 6.2*a*), principally by sulphate reduction, forms a layer of solid sulphides immediately below the surface of the sediment (see Figure 6.4*c*). During the winter, there is a net increase in the concentration of these sulphides as a result of relatively limited oxygen influx (Figure 6.2*b*). The situation in summer appears to change due to enhanced sediment irrigation by burrowing infauna. This supplies enough oxygen to not only oxidise the increased production (due to higher temperatures) of the reduced species (e.g. H_2S), but also oxidise the solid sulphides accumulated over the winter. The implication is that both oxic and anoxic bacterial reactions are present in spatially related micro-environments within the sediment column, and the degree of anoxia observed is the balance of the two.

6.3 Sulphur cycling

Figure 6.3 shows a mass balance model for sulphur speciations. Once other more thermodynamically amenable electron acceptors have been depleted, sulphate is used by sulphate reducing bacteria, such as *Desulfovibrio* sp. This occurs until the sulphate concentration falls to a point where it is outcompeted for the organic substrate by methanogenic bacteria. The sulphide produced by sulphate reduction is then removed or recycled by different processes depending on the environment of deposition.

Variations in the sulphate reduction rate have been shown to depend on four factors:

(a) temperature (which controls the rate of bacterial metabolism)
(b) pressure (which affects the removal of dissolved gases such as H_2S, CO_2)
(c) the concentration of sulphate
(d) the concentration of reactive carbon.

Since the work of Boudreau & Westrich (1984) and Kuivila *et al.* (1989) has shown that the sulphate concentration only becomes rate limiting at concentrations well below 10% that of sea water, and changes in pressure in tidal systems are fairly small, the two important controls on marine and

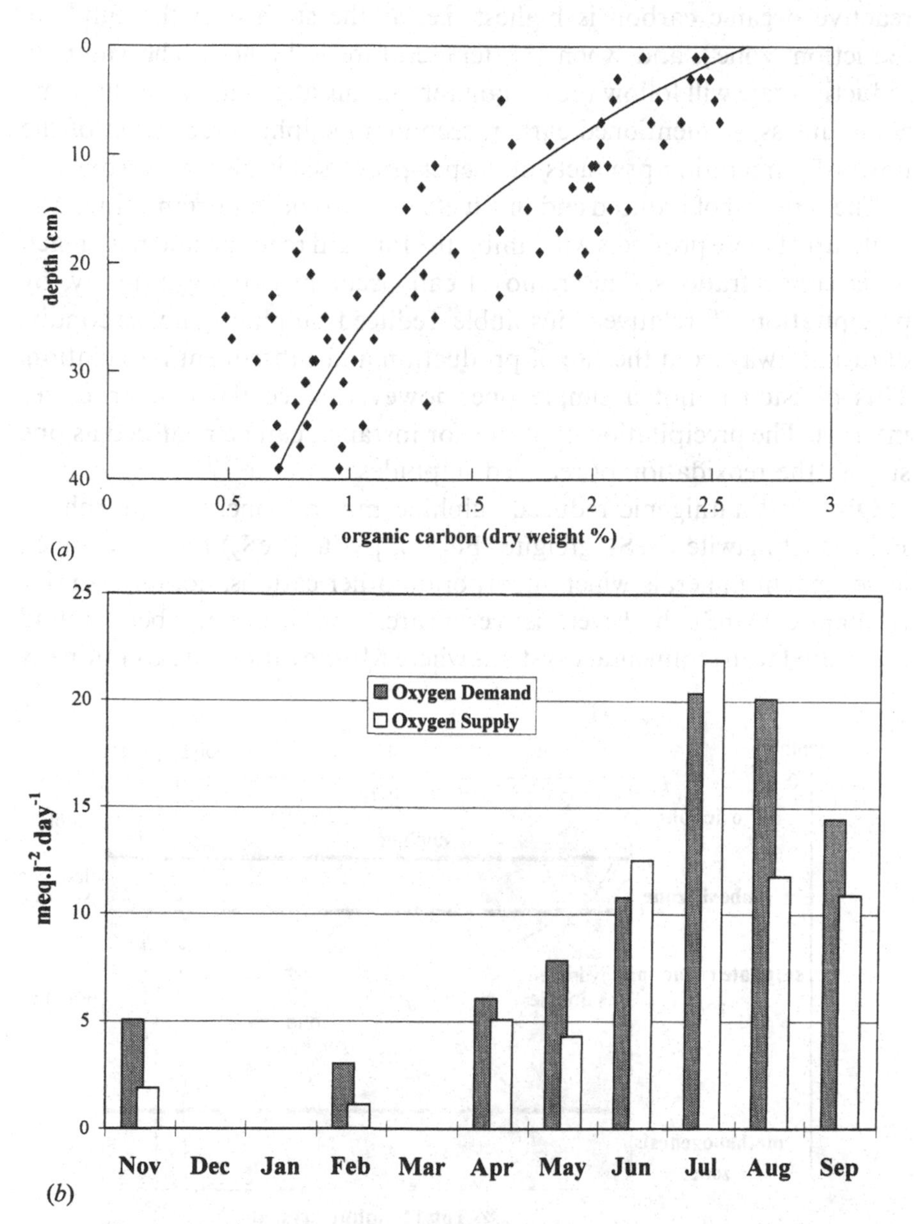

Figure 6.2 Carbon decomposition at Stiffkey, North Norfolk, UK (1990–1991) (Ruddy, 1993). (*a*) Organic carbon concentration with depth shows progressive steady state depletion due to decomposition. (*b*) Seasonal oxygen supply and demand estimated from modelling of carbon and sulphide oxidation rates.

intertidal sediment chemistry will be temperature, which will result in the seasonality of the cycle, and the reactive organic carbon concentration. Maximum rates of reduction, and therefore maximum rates of supply of reduced soluble sulphides and alkalinity, occur where the rate of supply of reactive organic carbon is highest (i.e. at the surface of the sulphate reduction 'zone'), and when the temperature is highest. The sulphate reduction rate will follow the carbon remineralisation rate over the same zone, unless, as mentioned earlier, secondary sulphate reduction of the upwardly migrating products of deeper processes is also occurring.

The removal of reduced end products will also be important, since any build up of these products will inhibit the forward reaction and may reach toxic concentrations. This removal can occur in two ways: firstly, by precipitation of relatively insoluble reduced sulphur, and, secondly, diffusion away from the site of production and subsequent reoxidation. This division is not a simple one, however, since the two processes interact. The precipitation of pyrite, for instance, can be regarded as one step in the reoxidation of reduced sulphides.

Observed authigenic reduced sulphide minerals include amorphous FeS, mackinawite (FeS), greigite (Fe_3S_4), pyrite (FeS_2) and to a much lesser extent minerals which incorporate other cations, notably Mn(II). Alabanite (MnS), however, is very rare, and has only been found associated with sedimentary systems where Mn/Fe ratios are exceptionally

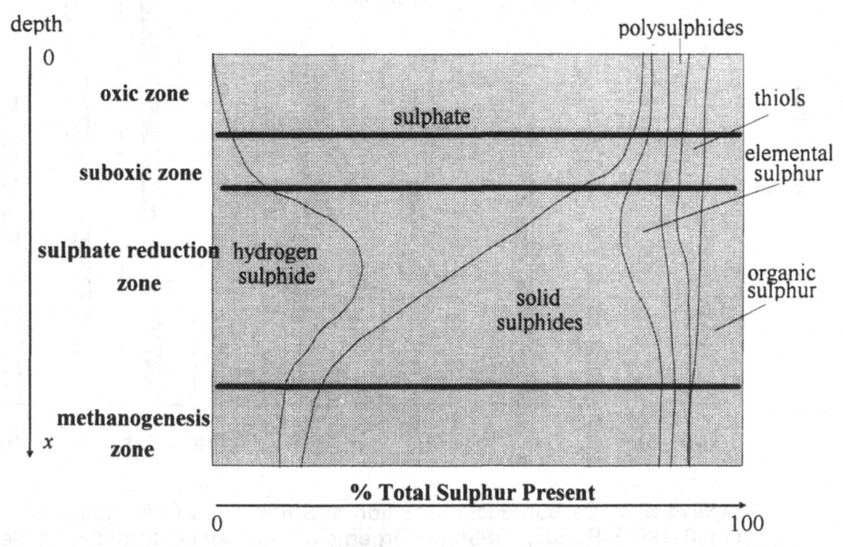

Figure 6.3 Notional sulphur speciation in sediments. Profiles of sulphur reservoirs observed in sediments, combined to give a qualitative picture of sulphur cycling processes.

high (e.g. in parts of the Baltic (Boesen & Postma, 1988)). Amorphous FeS, mackinawite and greigite are metastable (and more soluble) with respect to pyrite so they are generally not great repositories of reduced sulphur, whereas pyrite comprises the greater part of the total sulphur retained by the sediment.

The solubility products of the metastable iron sulphides are also not usually exceeded, except where sulphate reduction is at its peak, whereas pyrite is usually supersaturated in pore-waters throughout the year. This implies that to some extent it might form directly, without the precursor FeS thought to be necessary for framboid formation (cf. work by Luther *et al.*, 1992, discussed below). Lord & Church (1983) found that sulphide precipitation seemed to occur in two separate ways within the sulphate reduction zone. Pyrite concentrations, as single crystals rather than framboids, increased rapidly at the top of the sulphate reduction zone (to form about half of the pyrite buried), before sulphide became detectable in the pore-waters, whereas within the zone of pore-water sulphide accumulation, rates of formation were much slower. The sulphur in pyrite has an average oxidation state of − I, rather than − II, so an oxidising agent is obviously required at some stage in the process, and since it seems likely that pyritisation occurs by the addition of sulphur (rather than the removal of iron), it is reasonable to suggest that the oxidising agent might be an oxidised sulphur species such as elemental sulphur. Elemental sulphur often shows a broad peak just above the sulphate reduction zone, implying its formation by oxidation with either Fe(III) minerals, Mn(IV) minerals or indeed oxygen brought down by irrigation processes, but it is also found higher in the sediment profile, possibly associated with particular micro-environments. The labelled incubation studies of Howarth & Jorgensen (1984) show that sulphate reduction does occur within the oxic zone, further supporting this idea.

It may be that this elemental sulphur is the source of polysulphides (S_x^{2-}), which were proposed by Rickard (1975) to react with iron monosulphides and directly precipitate pyrite in euhedral form. This polysulphide is thought to result from the reaction of elemental sulphur and HS^- (Teder, 1971; Gigenbach, 1972). Luther (1991) suggests that the conversion from FeS to euhedral FeS_2 occurs via a complex formed by FeS and S_5^{2-}, since the step requires both FeS to be solubilised and a change in the electron configuration of the Fe(II). These laboratory studies seem to confirm Rickard's original mechanism.

The upper zone of sulphide precipitation can therefore be regarded as being formed by the reaction of FeS with elemental sulphur, which is known to nucleate FeS precipitation (Boulegue, Lord & Church, 1982), and is also thought to be available at the oxic-anoxic boundary. Since pore-water sulphide does not accumulate in this zone, the limiting factor on the formation of pyrite at this level would seem to be the availability of S(– II) rather than Fe(II) or S(0). In the lower zone, concentrations of metastable iron sulphides are low, implying rapid conversion to pyrite. It is therefore the formation of the metastable iron sulphides that becomes limiting to the process. Since dissolved sulphide is accumulating at this level, it is Fe(II) availability that is considered to be the limiting factor. The lower concentrations of more oxidised forms of sulphur (e.g. elemental sulphur) may also limit this process, since the zone is also characterised by greigite (Fe_3S_4), perhaps initially forming from iron monosulphide enriched pyrite – $Fe(II)_3$,S(0),$S(-II)_3$ – that eventually forms pyrite framboids by internal nucleation (Sweeney & Kaplan, 1973). The preservation of metastable iron sulphides can be expected where the availability of reduced sulphur is low, and they are often found associated with siderite cements, themselves apparently diagnostic of a relatively high Fe/S ratio (Pye *et al.*, 1990). It may be that the pyrite formed in this zone forms directly by reaction between H_2S_2 (itself formed by the reaction of S(– II) and S(0)) and Fe(II).

An alternative mechanism is suggested by Drobner *et al.* (1990), where the reaction between FeS and H_2S to produce pyrite and H_2 (where H^+ is the oxidant) was observed under anaerobic conditions in laboratory experiments. This ties in with the observation that H_2 might constitute an important source of energy for some bacterial communities (Crill & Martens, 1987).

The integrated sulphate reduction rates of coastal marine sediments far exceed the amount of reduced sulphur actually buried (e.g. Jorgensen, 1977), and assuming steady state this could occur either by loss of the sulphur in reduced form (e.g. the volatile gas dimethyl sulphide, DMS) or by loss of the sulphur during regeneration processes during mixing at the surface. The flux of reduced sulphides from the sediment by ebullition (transport of gas as bubbles), perhaps associated with fairly stable gas bubble tubes formed during the summer, was measured by Chanton, Martens & Goldhaber (1987) in shallow marine sediments, and was found to be a very small part of the sulphur budget. As far as regeneration of sulphate is concerned, sulphate/chloride ratios often exceed those of sea water in the oxic zone, which suggests that reoxidation may enhance sulphate concentrations and thereby total sulphate reduction. The reduced sulphur species involved may be either upwardly migrating sulphide ions or upwardly mixed (or stationary if the zones shift seasonally) solid iron sulphides, which are converted by both microbially mediated and abiotic oxidation to S(0), polysulphides and ultimately sulphate. Jorgensen (1990), using isotopically labelled incubations of fresh water sediments, showed that the majority of this anoxic sulphide oxidation proceeded by way of thiosulphate. Two-thirds of the initial product of this process was released into the pore-waters as thiosulphate, the remainder being sulphate. This thiosulphate was then transformed microbially into SO_4^{2-}, reduced to form S^{2-}, or disproportionated to produce SO_4^{2-} and S^{2-}. Overall, in this anoxic sediment, 72% of the thiosulphate sulphur became S^{2-} and the remainder SO_4^{2-}. Fossing & Jorgensen (1990) confirmed these findings on estuarine sediments, observing a very similar pattern.

The equilibrium of the resulting species and the reduced species they formed from will depend on the redox conditions at the time, and will probably involve those reactive iron(III) minerals that are present; hence the equilibrium is likely to vary from environment to environment. Davis-Colley, Nelson & Williamson (1985) have shown that this balance may control the behaviour of the sediment as a sink or source of pollutants since many form polysulphide-type complexes which can greatly enhance the solubility of some otherwise insoluble trace elements. Since pyrite is thought to be an important sink for trace metals (e.g. Raiswell & Berner, 1986; Morse & Cornwell, 1987), the cycling of these elements may be intimately linked to the cycling of sulphur. However, as Howarth & Jorgensen (1984) have pointed out, these relatively oxidised sulphur phases are very dynamic and can undergo a plethora of inorganic and organic redox reactions, usually in association with organic sulphides, so quantifying this part of the sulphur cycle is extremely difficult.

Luther *et al.* (1983) looked at organic sulphur cycling in saltmarsh pore-waters, and found that during cold periods organic sulphur compounds were nearly absent from the system. The large concentrations of tetrathionate in the oxic layer also suggested that much of this sulphur was converted to inorganic sulphur during the cold period. The most important organic sulphur species they found, which could reach concentrations equivalent to reduced inorganic sulphur in the sulphate reduction zone, was a thiol, possibly glutathione. These thiols are known to form under acid conditions by the reaction of organic molecules with sulphur anions, but they also form where pyrite production is at its maximum. This suggests that pyrite may be a starting point for their formation. Moreover, this reaction, by microbial carbon reduction, oxidised pyrite to form the thiol and sulphate in the absence of oxygen and without generating acidity, which fits the observations of relatively constant pH at this level in the sediment.

At the surface of some marine sediments, organic sulphur can comprise as much as 50% of the total sulphur present (François, 1987) due to biosynthesis which incorporates sulphur of all oxidation states, but also, because of the reactivity of sulphides and polysulphides, by chemical addition. There is usually an increasing S/C ratio with depth in sediments, partly associated with humic substances, and most of this increase occurs in the oxic and suboxic zones. This organic repository may be the source of the sulphur required to convert metastable iron sulphides, formed in the lower part of the sulphur reduction zone, to framboidal pyrite, which is often found closely associated with organic matter.

The availability of iron partly controls pyritisation processes, which in turn is an important control on the cycling of sulphur. Sulphur retention by the sediment system will depend on the presence of a layer of reactive iron oxides at the surface to precipitate upwardly migrating S(– II) (or available sulphides) which may in turn control trace metal mobilisation, sediment productivity and community metabolism. Free iron, which is the dominant cation in most of the diagenetic mineral phases, will likewise control the precipitation of environmentally diagnostic mineral assemblages, which suggests that, as an oxidant, Fe(III) accounts for a relatively large proportion of the cycling of electrons. Iron cycling may therefore constitute a major part of the budgeting of sulphur (and other elements) in coastal marine sediments. The processes involved are not well understood because of the complexity of the iron minerals and the poorly understood mechanisms of pyrite precipitation.

Luther *et al.* (1992) have suggested that in saltmarsh sediments, where primary productivity is very high, the production of organic ligands by plants and bacteria may form complexes with Fe(III) and Fe(II) and may therefore have a role to play in the cycling of iron. Their observations suggest that there are four important steps that account for its mass transfer. Firstly, the solubilisation of Fe(III) by organic ligands, which enhances this otherwise slow reaction, but on the other hand suggests a biotic seasonality. Secondly, this mobile oxidising agent will then be reduced by reduced species in the sediment, such as pyrite (thereby oxidising it). Thirdly, the reduced iron complex formed will be oxidised principally by Fe(III) minerals, which Luther *et al.* (1992) considered to be the dominant oxidant. The net effect of this is the transfer of electrons from the organic matter to Fe(III) minerals via a cyclic organo-Fe complex.

Where the sulphate reduction rate is high, producing an excess of H_2S, reduced iron will precipitate to give rise to a reduced sulphur pool, which comprises the fourth step. This reduced pool may then be reoxidised when H_2S levels fall off, either with depth or with seasonal change. Luther *et al.* (1992) considered that Mn(IV), being less abundant than Fe(III), did not form a significant reservoir for these electrons. In terms of electron flow, however, it should be noted that the size of the potential reservoir is only one of the important factors. The rate of the reactions involved and their free energy changes may offset any relative scarcity.

It is worth noting that although the specific mechanisms of the sulphur cycle are of critical importance to understanding marine-type sediment diagenesis, it is also true that the mechanisms themselves form a part of a wider dynamic system capable of responding to changes in conditions. For instance, the saltmarsh sediment at Stiffkey, North Norfolk, shows an interesting pattern of iron, manganese and sulphur accumulation with depth (Figure 6.4*a,b,c*) as well as seasonally. Pore-water concentrations of iron and manganese are high at the surface and during the winter, where the sediment is relatively anoxic. When oxygen supply is high relative to demand (i.e. below about 2 cm and during the summer), pore-water concentrations of these metals fall and solid sulphides are slowly oxidised. The net result of this is a system which preserved very little of its reducing potential (i.e. electrons). It results from rapid manganese and iron redox cycling (hourly and daily turnover of reservoirs) and longer term sulphur cycling (seasonal and yearly reservoir turnover), which form dynamic intermediate electron transfer cycles between the carbon source and the oxygen sink (Ruddy, 1993).

6.4 Conclusion

A summary of the diagenetic reactions concerning the transfer of electrons during burial is shown in Figure 6.5. Each line represents a flux of electrons from one reservoir to another. Each environment will show a different balance of these reservoirs, since each biodiagenetic regime will be dominated by different fluxes used to accommodate the flow of electrons from the organic matter. Ultimately, these electrons will be either buried, as authigenic minerals or as geopolymers, or leave the system by reaction with oxygen. It is well established that even in the most reducing environments (i.e. where anaerobic metabolism accounts for the bulk of the carbon degradation), only a small proportion of the electron flux that leaves the organic material is actually buried (principally as

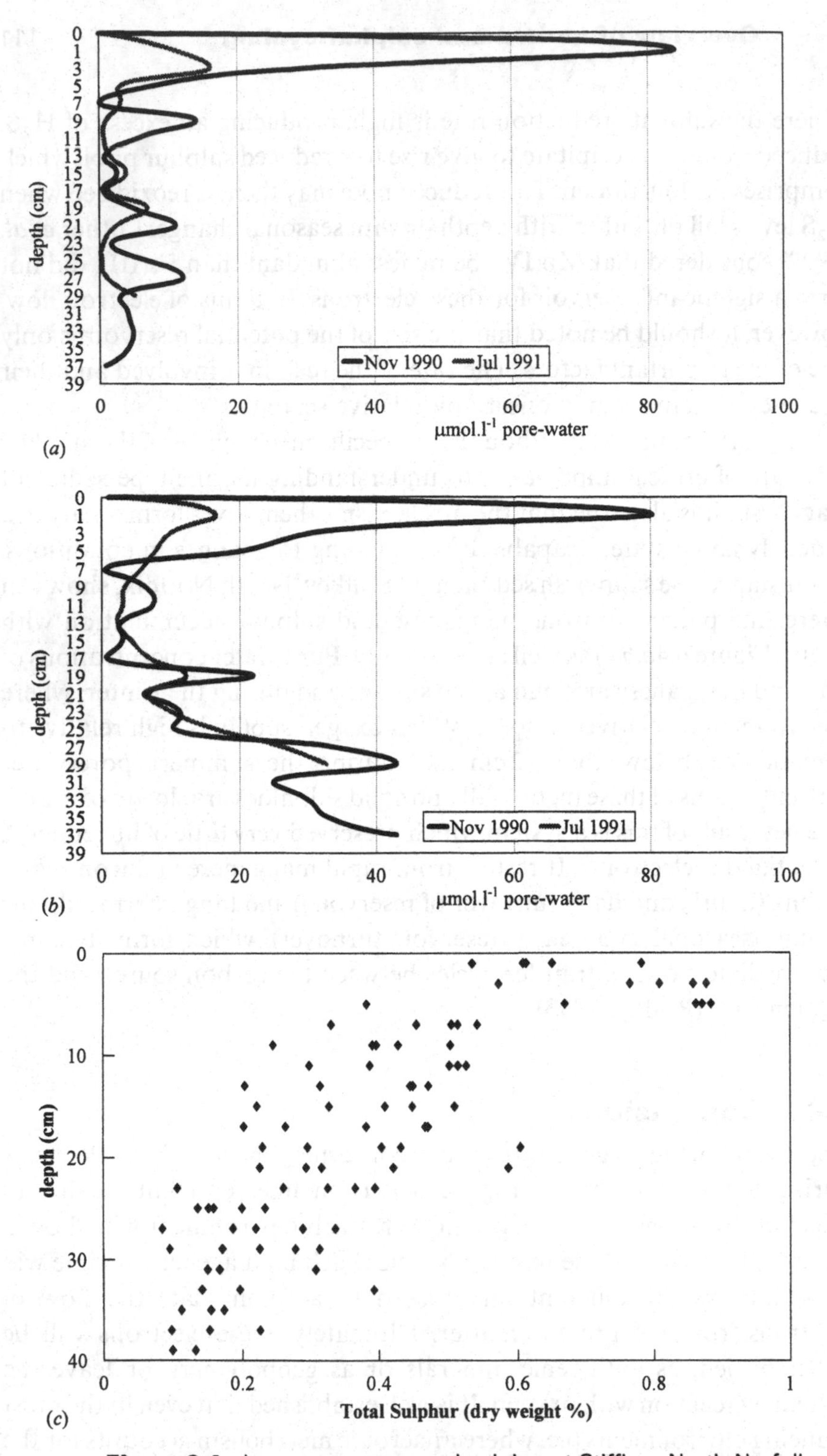

Figure 6.4 Iron, manganese and sulphur at Stiffkey, North Norfolk (1990–1991) (Ruddy, 1993). (*a*) Pore-water iron profiles. (*b*) Pore-water manganese profiles. (*c*) Solid sulphide (predominantly pyrite) profile.

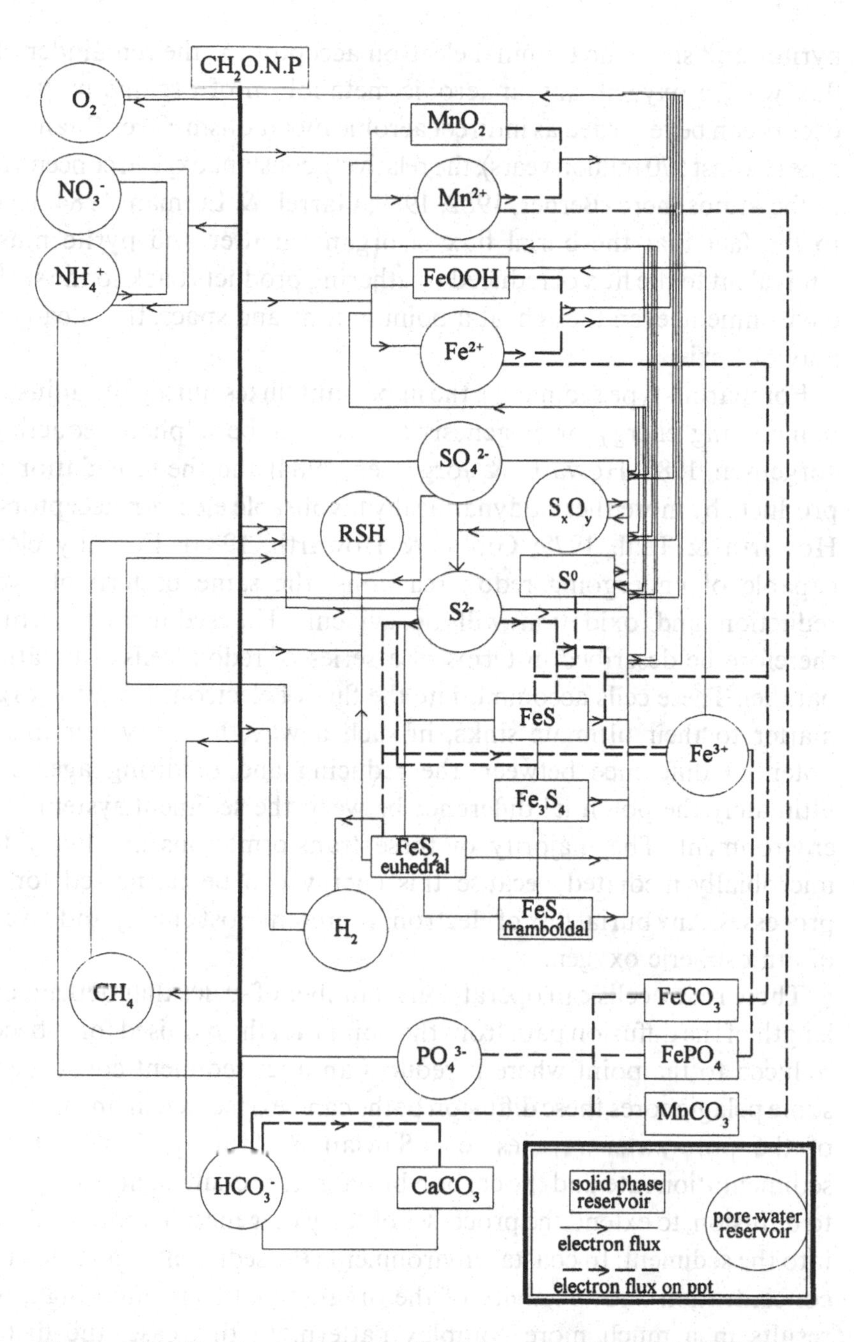

Figure 6.5 Electron transfer reactions for sediments (Ruddy, 1993). $CH_2O.N.P$ (organic matter) is transformed by bacterial decomposition reactions to bicarbonate in the pore-waters. This is the primary source of electrons (and therefore energy) for the remainder of the sediment redox chemistry. Most of the primary flux of electrons may pass through the sulphur, iron and manganese cycles, but will eventually react with oxygen. Only a small part of the total electron flux will ultimately be buried as reduced minerals.

pyrite), and since the terminal electron acceptor for the remainder of the flux will be oxygen, any anaerobic metabolism of organic matter that occurs can be regarded as indirect aerobic metabolism. Over Phanerozoic time (the last 570 million years), the relatively constant oxygen concentration in the atmosphere (Berner, 1982, 1989; Garrels & Lerman, 1984) testifies to the fact that the burial flux of organic matter and pyrite must be equivalent to the flux of reduced weathering products back to the surficial environment, even though, at a point in time and space, this steady state may not exist.

For marine-type sediments, the important fluxes during the adjustment of incoming 'energy' or diagenesis, are likely to be sulphate reduction (cf. Jorgensen, 1982; Howarth & Jorgensen, 1984) and the reoxidation of its products by more thermodynamically favourable electron acceptors (e.g. Howarth & Teal, 1979; Giblin & Howarth, 1984). For any element capable of undergoing redox reactions, the same pattern of cyclical reduction and oxidation will be evident. The sediment system can therefore be described in terms of a series of redox 'cells', operating in parallel. These cells accommodate the flux of electrons from the organic matter to their ultimate sinks, in such a way that they minimise the potential difference between the reducing and oxidising agents, and ultimately the potential difference between the sediment system and its environment. The majority of these transformations are likely to be microbially mediated because this energy can be harnessed for their processes. Any burial flux of electrons represents lost energy and an excess of atmospheric oxygen.

These redox cells can operate on a number of scales that depend on the length of the diffusion path from the point that the oxidised form becomes reduced to the point where it reduces another sediment constituent. In some pelagic cores these diffusion paths can be observed in linear portions of the pore-water profiles (e.g. Sawlan & Murray, 1983). Here the sedimentation rate and the carbon burial rate are sufficiently low, relative to diffusion, to extend the processes of early diagenesis over tens of metres into the sediment. In coastal environments the sedimentation rate and the concentration and reactivity of the organic matter is often high, which results in a much more complex pattern. In this case, the distances between the cells are much shorter, since by definition the adjustment must occur more rapidly. Like laminar and turbulent flow, there may come a point where the flow of electrons downwards is better dispersed through 'eddies', which in this case are transitory micro-environments with small-scale three dimensional diffusion, rather than more stable zones separated by large scale one dimensional diffusion. The ubiquitous nature of bacteria ensures that the microbial ecology at a point in the sediment can keep pace with the changing conditions, so one can envisage centres of primary microbial activity surrounded by areas of redox transfer in a shifting pattern with time (and therefore depth). At a point in time, therefore, the 'zones' of electron acceptor use will be arranged radially about these centres of activity, with the redox cells transferring the electrons outwards towards areas where the most thermodynamically favoured electron acceptors are still relatively available.

The important point is that no redox reaction operates in isolation; they form chains of electron transfer that release energy and drive sediment diagenesis. The bulk chemistry observed during sampling may mask the very dynamic nature of some of the elements, since a sample will 'average out' the differences that may be present over very small distances. Connections between different parts of sediment diagenesis (e.g. between microbial redox reactions and authigenic minerals), that may be very direct at the microscale, may not therefore be apparent. Conversely, factors such as bioturbation and bioirrigation may determine the length of diffusion paths between redox cells, and these need to be sampled on a larger scale.

Acknowledgements

This work was funded by a NERC PhD studentship at the University of East Anglia, Norwich, UK. I should like to thank Julian Andrews, Tim Jickells and Carol Turley for critical reviews of the manuscript.

References

Abdollahi, H. & Nedwell, D.B. (1979) Seasonal temperature as a factor influencing bacterial sulfate reduction in a salt marsh. *Microb. Ecol.* **5**, 73–9.

Aller, R.C. (1980) Diagenetic processes near the sediment–water interface of Long Island Sound 1. Decomposition and nutrient element in geochemistry (S, N, P). In *Estuarine Physics and Chemistry: Studies in Long Island Sound* (ed. B. Saltzman), *Adv. in Geophysics* **22**, 238–350. Academic Press, New York.

Aller, R.C. (1982) The effects of macrobenthos on the chemical properties of marine sediments and overlying water. In *Animal–Sediment Relations: The Biogenic Alteration of Sediments* (eds. P.L. McCall & M.J.S. Tevesz), Plenum Press, New York, 53–102.

Aller, R.C. (1984) The importance of relict burrow structures and burrow irrigation in controlling sedimentary solute distributions. *Geochim. Cosmochim. Acta* **48**, 1929–34.

Berner, R.A. (1977) Stoichiometric models for nutrient regeneration in anoxic sediments. *Limnol. Oceanog.* **22**, 781–6.

Berner, R.A. (1978) Sulfate reduction and the rate of deposition of marine sediments. *Earth Planet. Sci. Letters* **37**, 492–8.

Berner, R.A. (1982) Burial of organic matter and pyrite sulfur in the modern ocean: its geochemical and environmental significance. *Am. Jour. Sci.* **282**, 451–73.

Berner, R.A. (1989) Biogeochemical cycles of carbon and sulfur and their effect on atmospheric oxygen over Phanerozoic time. *Palaeogeog. Palaeoclim. Palaeoecol.* (Global & Planetary change section) **75**, 95–122.

Berner, R.A. & Westrich, J.T. (1985) Bioturbation and the early diagenesis of carbon and sulfur. *Am. Jour. Sci.* **285**, 193–206.

Boesen, C. & Postma, D. (1988) Pyrite formation in anoxic environments of the Baltic. *Am. Jour. Sci.* **288**, 575–603.

Boudreau, B.P. & Westrich, J.T. (1984) The dependence of bacterial sulfate reduction on sulfate concentration in marine sediments. *Geochim. Cosmochim. Acta* **48**, 2503–16.

Boulegue, J., Lord, C.J. & Church, T.M. (1982) Sulfur speciation and associated trace metals (Fe, Cu) in the pore waters of Great Marsh, Delaware. *Geochim. Cosmochim. Acta* **46**, 453–64.

Chanton, J.P., Martens, C.S. & Goldhaber, M.B. (1987) Biogeochemical cycling in an organic-rich coastal marine basin. 7. Sulfur mass balance, oxygen uptake and sulfide retention. *Geochim. Cosmochim. Acta* **51**, 1187–99.

Claypool, G. & Kaplan, I.R. (1974) The origin and distribution of methane in marine sediments. In *Natural Gases in Marine Sediments* (ed. I.R. Kaplan), Plenum Press, New York, 99–139.

Coleman, M.L. (1985) Geochemistry of diagenetic non-silicate minerals: kinetic considerations. *Phil. Trans. R. Soc. London* A **315**, 39–56.

Crill, P.M. & Martens, C.S. (1986) Methane production from bicarbonate and acetate in an anoxic marine sediment. *Geochim. Cosmochim. Acta* **50**, 2089–97.

Crill, P.M. & Martens, C.S. (1987) Biogeochemical cycling in an organic-rich coastal marine basin. 6. Temporal and spatial variations in sulfate reduction rates. *Geochim. Cosmochim. Acta* **51**, 1175–86.

Davis-Colley, R.J, Nelson, R.D. & Williamson, K.J. (1985) Sulfide control of cadmium and copper concentrations in anaerobic estuarine sediments. *Mar. Chem.* **16**, 173–86.

Devol, A., Anderson, J., Kuivila, K. & Murray, J. (1984) A model for coupled sulfate reduction and methane oxidation in the sediments of Saanich Inlet. *Geochim. Cosmochim. Acta* **48**, 993–1004.

Drobner, E., Huber, H., Wachtershauer, G., Rose, D. & Stetter, K.O. (1990) Pyrite formation linked with hydrogen evolution under anaerobic conditions. *Nature* **346**, 742–4.

Fossing, H. & Jorgensen, B.B. (1990) Oxidation and reduction of radiolabelled inorganic sulfur compounds in an estuarine sediment, Kysing fjord, Denmark. *Geochim. Cosmochim. Acta* **54**, 2731–42.

François, R. (1987) A study of sulfur enrichment in the humic fraction of marine sediments during early diagenesis. *Geochim. Cosmochim. Acta* **51**, 17–27.

Froelich, P., Klinkhammer, G., Bender, M., Luedtke, N., Heath, G., Cullen, D., Dauphin, P., Hammond, D., Hartman, B. & Maynard, V. (1979) Early oxidation of organic matter in pelagic sediments of the eastern equatorial Atlantic: suboxic diagenesis. *Geochim. Cosmochim. Acta* **41**, 1075–90.

Garrels, R.M. & Lerman, A. (1984) Coupling of the sedimentary sulfur and carbon cycles – an improved model. *Am. Jour. Sci.* **284**, 989–1007.

Giblin, A.E. & Howarth, R.W. (1984) Pore water evidence for a dynamic sedimentary iron cycle in salt marshes. *Limnol. Oceanog.* **29**, 47–63.

Gigenbach, W. (1972) Optical spectra and equilibrium distribution of polysulfide ions in aqueous solution at 20°C. *Inorg. Chem.* **11**, 1724–30.

Goldhaber, M. & Kaplan, I. (1974) The sulfur cycle. In *The Sea, Vol. 5* (ed. E.D. Goldberg). Wiley, New York, 569–655.

Hines, M.E. & Jones, G.E. (1985) Microbial biogeochemistry and bioturbation in the sediments of Great Bay, New Hampshire. *Estuar. Coast. Sh. Sci.* **20**, 729–42.

Howarth, R.W. & Jorgensen, B.B. (1984) Formation of ^{35}S labelled elemental sulfur and pyrite in coastal marine sediments (Limfjorden, Denmark) during short term $^{35}SO_4$ reduction measurements. *Geochim. Cosmochim. Acta* **48**, 1807–18.

Howarth, R.W. & Teal, J.M. (1979) Sulfate reduction in a New England salt marsh. *Limnol. Oceanog.* **24**, 999–1013.

Howes, B.L., Dacey, J.W.H. & King, G.M. (1984) Carbon flow through oxygen and sulfate reduction pathways in salt marsh sediments. *Limnol. Oceanog.* **29**, 1037–51.

Janke, R., Emerson, S. & Murray, J. (1982) A model of oxygen reduction, denitrification and organic matter mineralization in marine sediments. *Limnol. Oceanog.* **27**, 610–23.

Jorgensen, B.B. (1977) The sulfur cycle of a coastal marine sediment (Limfjorden, Denmark). *Limnol. Oceanog.* **5**, 814–32.

Jorgensen, B.B. (1982) The ecology of the bacteria of the sulfur cycle with special reference to anoxic–oxic interface environments. *Phil. Trans. R. Soc. London* B **298**, 543–61.

Jorgensen, B.B. (1990) The sulfur cycle in freshwater sediments: role of thiosulfate. *Limnol. Oceanog.* **35**, 1329–42.

Kuivila, K., Murray, J. & Devol, A. (1989) Methane production, sulfate reduction and competition for substrates in the sediments of Lake Washington. *Geochim. Cosmochim. Acta* **53**, 409–16.

Lerman, A. (1979) *Geochemical Processes: Water and Sediment Environments*, Wiley, New York.

Li, Y-H. & Gregory, S. (1974) Diffusion of ions in sea water and deep-sea sediments. *Geochim. Cosmochim. Acta* **38**, 703–14.

Lord, C.J. & Church, T.M. (1983) The geochemistry of salt marshes: sedimentary ion diffusion, sulfate reduction, and pyritization. *Geochim. Cosmochim. Acta* **47**, 1381–91.

Luther, G.W. (1991) Pyrite synthesis via polysulfide compounds. *Geochim. Cosmochim. Acta* **55**, 2839–49.

Luther, G.W., Church, T.M., Scudlark, J.R. & Cosman, M. (1983) Inorganic and organic sulfur cycling in salt marsh pore waters. *Science* **232**, 746–9.

Luther, G.W., Kostka, J.E., Church, T.M., Sulzberger, B. & Stumm, W. (1992) Seasonal iron cycling in the salt-marsh sedimentary environment: the importance of ligand complexes with Fe (II) and Fe (III) in the dissolution of Fe (III) minerals and pyrite, respectively. *Mar. Chem.* **40**, 81–103.

Martens, C.S. & Klump, J.V. (1984) Biogeochemical cycling in an organic-rich coastal marine basin. 4. An organic carbon budget for sediments dominated by sulfate reduction and methanogenesis. *Geochim. Cosmochim. Acta* **48**, 1987–2004.

McNichol, A., Lee, C. & Druffel, R. (1988) Carbon cycling in coastal sediments: 1. A quantitative estimate of the remineralization of organic carbon in the sediments of Buzzards Bay, MA. *Geochim. Cosmochim. Acta* **52**, 1531–43.

Middleburg, J.J. (1989) A simple rate model for organic matter decomposition in marine sediments. *Geochim. Cosmochim. Acta* **53**, 1577–81.

Morse, J. & Cornwell, J. (1987) Analysis and distribution of iron sulfide minerals in recent anoxic marine sediments. *Mar. Chem.* **22**, 55–69.

Pye, K., Dickson, J., Schiavon, N., Coleman, M. & Cox, M. (1990) Formation of siderite-Mg-calcite-iron concretions in intertidal marsh and sandflat sediments, north Norfolk, England. *Sedimentology* **37**, 325–43.

Raiswell, R. & Berner, R.A. (1986) Pyrite and organic matter in Phanerozoic normal marine shales. *Geochim. Cosmochim. Acta* **50**, 1967–76.

Richards, F.A., Cline, J.D., Broenkow, W.W. & Atkinson, L.P. (1965) Some consequences of the decomposition of organic matter in Lake Nitinat, an anoxic fjord. *Limnol. Oceanog. Suppl.* **10**, R185–R201.

Rickard, D.T. (1975) Kinetics and mechanism of pyrite formation at low temperatures. *Am. Jour. Sci.* **275**, 636–52.

Ruddy, G. (1993) Microenvironmental modelling of redox chemistry in salt marsh sediments. PhD Thesis. University of East Anglia, Norwich, UK.

Sawlan, W. & Murray, J.W. (1983) Trace metal remobilization in the interstitial waters of red clay and hemipelagic marine sediments. *Earth Planet. Sci. Letters* **64**, 213–30.

Sharma, P., Gardner, L., Moore, W. & Bollinger, M. (1987) Sedimentation and bioturbation in a salt marsh as revealed by ^{210}Pb, ^{137}Cs and ^{7}Be studies. *Limnol. Oceanog.* **32**, 313–26.

Stumm, W. & Morgan, J.J. (1970) *Aquatic Chemistry*, Wiley, New York.

Sweeney, R.E. & Kaplan, I.R. (1973) Pyrite framboid formation: laboratory synthesis and marine sediments. *Econ. Geol.* **68**, 618–34.

Swider, K.T. & Makin, J.E. (1989) Transformations of sulfur compounds in marsh-flat sediments. *Geochim. Cosmochim. Acta* **53**, 2311–23.

Teder, A. (1971) The equilibrium between elementary sulfur and aqueous polysulfide solutions. *Acta Chem. Scand.* **25**, 1711–28.

Troelson, H. & Jorgensen, B.B. (1982) Seasonal dynamics of elemental sulfur in two coastal marine sediments. *Estuar. Coast. Sh. Sci.* **15**, 255–66.

Welte, D. (1973) Recent advances in organic geochemistry of humic substances and kerogen. A review in *Advances in Organic Geochemistry* (ed. J.P. Riley and F. Bienner). Editions Technip, Paris, 4–13.

7

○ ○ ○ ○ ○ ○ ○ ○ ○ ○ ○ ○ ○ ○ ○ ○ ○ ○ ○ ○

Microbial activity and diagenesis in saltmarsh sediments, North Norfolk, England

K. Pye, M.L. Coleman and W.M. Duan

7.1 Introduction

The relationship between microbial processes, diagenetic reactions and mineralisation within intertidal sediments, including the development of diagenetic carbonate concretions, tubules and localised cemented layers, has attracted major interest in recent years. Such phenomena are relatively common in intertidal sediments around the British Isles and on other European coasts (see e.g. Pye, 1981; Andrews, 1988; Al-Agha *et al.*, 1995; Duck, 1995). The processes associated with their formation are important in relation to the cycling of carbon, sulphur and iron, as well as the fate of industrial and other pollutants. Early diagenesis within saltmarshes in particular is of importance because of the high biological productivity which is characteristic of such environments, and their position at the interface between marine and terrestrial systems (see Vernberg, 1993, for a recent review of saltmarsh processes).

The role of microbial activity in early diagenesis has been recognised for some time, and a variety of approaches has been used to investigate these processes, including, for example, studies of C–S–Fe interactions (Berner & Westrich, 1985), rates of sulphate reduction (Jorgensen, 1978; Jorgensen & Bak, 1991), nitrate reduction (Sorenson, 1987) and Fe/Mn reduction (Lovley, 1991). Recently, significant progress has been made in quantifying the rates of microbial reactions and the interactions between biotic and abiotic factors (e.g. Blackburn & Blackburn, 1993; Parkes *et al.*, 1993a,b), partly as a result of the application of biochemical techniques which allow detailed specification of microbial biomass, community structure and nutritional status (e.g. White, 1993; Eglinton, Parkes & Zhao, 1993). Important advances in understanding have also been made as a result of experimental investigations which have sought to define the effects of various limiting environmental factors on rates of microbial activity (e.g.

Lovley, 1991; Coleman *et al.*, 1993), and through the development of improved field methods of micro-scale sampling and subsequent laboratory analysis (e.g. Davison *et al.*, 1991).

This contribution provides an illustration of the application of biochemical and geochemical techniques in the context of an ongoing investigation into the inter-relationships between microbial activity and sediment diagenesis within saltmarsh and intertidal flat sediments near Warham on the north Norfolk coast, eastern England (Figure 7.1). This investigation,

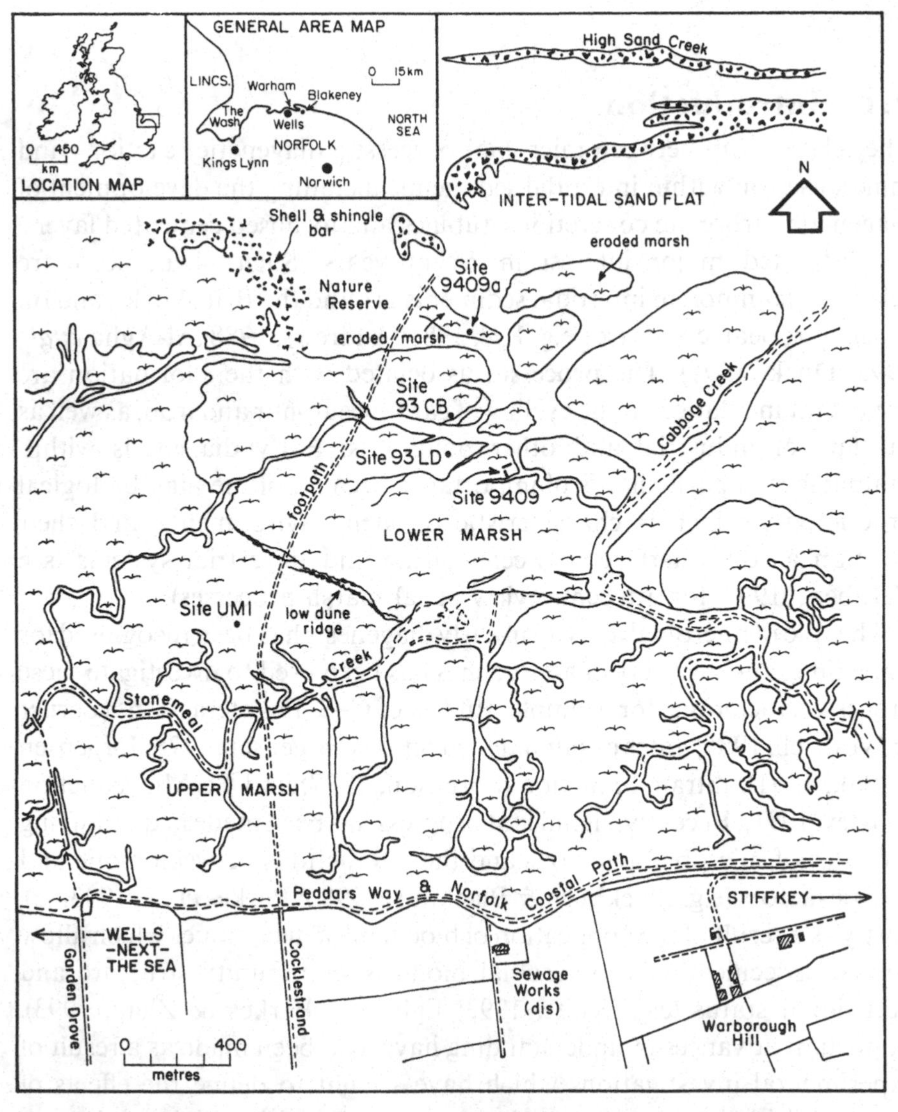

Figure 7.1 Location of the study area and sampling locations.

which involves inter-disciplinary collaboration between complementary research teams at the Universities of Reading, Tennessee, Leeds and the United States Geological Survey, uses a number of investigative approaches, including studies of solid phase mineralogy, texture and chemistry, pore-water sampling and analysis, laboratory and field experimentation, and characterisation of recent microbial activity through the application of signature lipid biomarker techniques. This contribution is concerned mainly with the information provided by biomarker and associated geochemical analysis of sediments from two marshes of differing age and recent sedimentation history. The general nature of the information which can be provided by biomarkers is briefly reviewed below, but for more detailed information the reader should refer to other recent reviews by White (1993) and Eglinton *et al.* (1993).

7.2 Nature of the study area

The Warham saltmarshes form part of an extensive belt which extends for a distance of 35 km along the coast of north Norfolk. The general setting and geomorphological character of the marshes have previously been reviewed by Pye (1992). At Warham there are two distinctive marsh morpho-stratigraphic units, the Upper Marsh and the Lower Marsh, to seaward of which lies a wide intertidal sand flat. The surface of the Upper Marsh lies at *c.* 2.90 m O.D. (Ordnance Datum) at its seaward end, declining to 2.73 m O.D. near the landward limit. The level of mean high water spring tides (MHWST) decreases eastwards along the coast from 3.00 m O.D. at Wells Bar, approximately 3 km to the west of the study site, to 2.65 m O.D. at Blakeney Bar, approximately 4 km to the east. The respective heights for the predicted Highest Astronomical Tide are 3.10 m and 2.75 m O.D. Consequently the Warham Upper Marsh is covered only by the highest normal spring tides and periodic storm surges (Pye, 1992). The Upper Marsh is separated from the Lower Marsh by a low dune ridge which has a crest height of 3.5 to 3.6 m O.D. and a maximum width of *c.* 20 m. Levees, which have a maximum elevation of 3.4 m, but more typically lie in the range 3.1 to 3.2 m O.D., occur along the banks of the major transverse creek (Stonemeal Creek) which separates the Upper Marsh into inner and outer sections. Smaller levees flank the secondary creeks, forming discrete areas of low-lying backmarsh which is prone to waterlogging after high tides or heavy rain. Linear salt pans, which represent the abandoned remnants of former tidal creeks, are numerous (Figure 7.2). The remaining active creeks show a high degree of sinuosity and lateral stability.

Drilling investigations, palaeoecological analyses and the results of radiocarbon dating have shown that marshes began to form along the coastal fringe between Morston and Wells at least 6600 ^{14}C yr ago (Funnell & Pearson, 1989). At Stiffkey, located just to the east of the Warham marshes, sediments interpreted as saltmarsh deposits extend to a depth of approximately − 7 m O.D. Archaeological remains of Romano-British age have been identified on several parts of the inner marshes along the coast, testifying to an age of > 2000 years. Studies at Warham have also shown that industrial pollutant metals are restricted to the uppermost 5–10 cm of the sediment sequence on the Upper Marsh (Pye *et*

Figure 7.2 Oblique air photograph of the Warham Upper Marsh, showing the main transverse creek (Stonemeal Creek) and the Cocklestrand path (running left to right across the lower part of the photograph).

al., unpublished data), indicating only 5–10 cm of accretion in the past 100–150 years. This marsh can be regarded as mature in the sense that it has achieved a surface level which is in equilibrium with the local tidal frame. In the context of the microbial and geochemical processes considered here, its sedimentological and geomorphological regime can be regarded as essentially 'steady'.

The upper 2 m of sediments on the Upper Marsh consist mainly of fine silts with localised sandier facies along the courses of the larger tidal creeks (both active and inactive). The surface is generally well-vegetated by halophytic plants. The higher, better drained areas along creek levees and the margins of the seaward dune ridge are vegetated chiefly by *Halimione portulacoides*, *Sueda* spp., *Puccinellia maritima* and other grass species such as *Festuca* spp., and *Juncus* spp. Areas of intermediate elevation are vegetated mainly by *Aster tripolium*, *Limonium vulgare*, *Halimione portulacoides* and *Salicornia perennis*. Waterlogged areas and the margins of some salt pans are often characterised by the presence of patches of *Salicornia* spp. and *Spartina* spp. The vegetation forms a continuous sward with a dense root mat in the uppermost 20 cm of sediment. During the eighteenth and nineteenth centuries attempts were made to enclose part of the marshes between Warham and Wells, just to the west of the study area, but these attempts failed and were abandoned (Purchas, 1965). The marshes at Warham itself have never been embanked and have been modified by man-made drainage diversions to a relatively minor degree. In recent decades the marsh has experienced only very light to no grazing pressure and its principal usage has been for wildfowling and as a nature reserve. Between the late 1930s and early 1950s the marshes and sandflats to seaward were used for military target practice with firing positions located at Warham and Stiffkey.

The Warham Lower Marsh is a much younger feature which developed rapidly after the Second World War, and particularly from the mid 1950s onwards, when vegetation colonised the higher part of the intertidal sand flats to seaward of the dune ridge that forms the seaward boundary of the Upper Marsh. During approximately the past 45 years the rate of vertical accretion on the Lower Marsh has been rapid, averaging *c.* 1.5 cm y^{-1}. The landward part of the Lower Marsh currently has a surface elevation of 2.30 to 2.35 m O.D., falling to *c.* 2.20 to 2.25 m O.D. near the seaward edge. At present the marsh is covered by approximately 20% of all high tides. The vegetation cover on the Lower Marsh is dominated by lower saltmarsh communities which include species such as *Aster*, *Spartina*, *Limonium* and *Halimione* spp. Weakly developed levees are present along some of the larger creeks. Much of the intervening marsh is poorly drained and remains waterlogged for considerable periods after tidal flooding. Channel pans are less common than on the Upper Marsh and the density of active tidal creeks is higher. Particularly in the outer part of the Lower Marsh, many of the tidal creeks display an anastomising character and have frequently shifted their lateral position. The seaward margin of the marsh has suffered episodic wave erosion in the past ten years, forming areas of devegetated but root-bound muddy sand which have been partially buried by landward moving sand lobes originating from the intertidal flats. The intertidal flats themselves have an elevation ranging from 1.5 to 1.9 m O.D. Higher areas are represented by sand bars and localised accumulations of shingle and shells (mainly *Cerastoderma*, *Ostrea* and *Mytilus* spp.).

Within the Lower Marsh sediment sequence there is a sharp texture boundary at a depth of 45–70 cm, depending on the surface elevation (Figure 7.3). The sudden change from sandy muds to relatively clean, medium sands reflects the onset of marsh vegetation colonisation and mud accumulation during the early 1950s. A permanent groundwater table is present at a depth of approximately 0.8–0.9 m, although its precise level varies diurnally and seasonally in response to tidal forcing and weather conditions (Allison & Pye, 1994). Carbonate concretions, many of which contain shrapnel and other debris, are widespread in the sediments close to the water table and occasionally at other levels (Figure 7.3). The majority of the concretions, which may attain a maximum dimension of 25 cm, are cemented mainly by siderite, Mg-calcite and iron monosulphide (Pye, 1981, 1984; Pye *et al.*, 1990). Localised beds of siderite and siderite–goethite cemented root tubules, together with patches of cemented creek bed sediments, are also found. Field experiments on the Lower Marsh showed that small concretions and coatings can form around various nucleii within as little as 12 months (Pye *et al.*, 1990; Allison & Pye, 1994). Biogenic remains within the concretions have been extensively replaced by authigenic minerals. Replacement of plant material includes (a) coating and replacement of cells walls only, (b) wall replacement by siderite and cell infilling with siderite or iron monosulphide, and (c) infilling of roots and stem moulds with acicular and/or botryoidal siderite. The occurrence of siderite within the concretions and bioclast coatings is particularly interesting. Previous studies have associated the formation of siderite with the degradation of organic matter by anaerobic, methanogenic bacteria, typically in fresh water, brackish or deep-burial environments where limited supply of sulphate restricts production of H_2S by sulphate-reducing bacteria (SRB).

However, the results of water chemistry and solid phase stable isotope analyses indicate that the Warham siderite is forming under very different conditions. Analysis of pore-waters and surface waters has demonstrated that fresh water dilution is very localised, being restricted mainly to zones of groundwater seepage. The majority of concretions occur in sediments whose pore-waters have a similar composition to sea water, or which are slightly supersaline due to evapotranspiration (Pye *et al.*, 1990; Allison & Pye, 1994). The siderite concretions have markedly negative $\delta^{13}C$ values, averaging − 5.9‰ PDB, suggesting that the siderite is not formed under methanogenic conditions. Bacterial assays have shown that SRB are abundant and have high activities in the Lower Marsh, and siderite formation is unlikely to be related to a shortage of H_2S.

Figure 7.3 A typical section through the Lower Marsh, showing the upper laminated mud unit, the intermediate zone of brownish sands, and the lower zone of black, monosulphide stained sands containing an *in situ* concretion.

Concretions similar to those found in the Lower Marsh sediments occur locally along the banks and beds of creeks on the Upper Marsh, and in some saltmarsh pans, but are otherwise absent. Both *in situ* and reworked concretions occur in the intertidal flat sediments just beyond the seaward edge of the Lower Marsh. The reworked concretions are often partly oxidised and contain a range of secondary minerals including akaganeite, gypsum and goethite (Pye, 1988).

The contrasting ages and sedimentation histories of the two Warham marshes provide a useful framework for a comparative investigation of the geochemical environmental conditions prevailing and their relationship to microbial processes and mineral authigenesis. With this in mind, several trenches and pits were dug on the Upper Marsh and Lower Marsh during the period March 1993 – May 1995 (Figure 7.4). Samples of

Figure 7.4 Trench section at Site 9409 on the Lower Marsh.

sediment were collected at closely spaced depth intervals for examination of the solid phase mineralogy and chemistry, microbial activity and community structure, and pore-water composition. Detailed discussion of the results will be presented elsewhere, but a summary of the analytical approach and some typical results are presented below.

7.3 Lipid biomarkers as indicators of microbial activity and community structure

The signature lipid biomarker (SLB) technique is based on the premise that all micro-organisms have cell membranes composed of lipids, in particular phospholipids. By extracting and characterising the fatty acids associated with these phospholipids (PLFA), an assessment of the viable microbial biomass, community structure and nutritional status can be made. Owing to the labile nature of phospholipids, viable microbial biomass can be determined by quantifying the total concentration of PLFA in a sample. Patterns of PLFA recovered from individual isolates of bacteria can be used to characterise and identify bacteria at the species level and to estimate community compositions (Tunlid & White, 1992). Ratios of specific PLFA have also been demonstrated to relate to the nutritional or metabolic status of certain bacteria (Guckert *et al.*, 1985). In the past decade the SLB technique has been applied in a variety of marine and estuarine contexts (e.g. Baird *et al.*, 1985; Findlay *et al.*, 1990; Parkes *et al.*, 1993a,b), but the work outlined here represents the first known attempt to apply it in the context of saltmarshes.

7.4 Microbial activity in the Warham Upper Marsh

7.4.1 Investigative methods

A 1.55 m deep trench, measuring 3 m long and 1 m wide, was dug in the outer part of the Upper Marsh in October 1993 (site UM1 in Figure 7.1). Sediment samples were collected at 3–5 cm intervals down to the water table for determination of grain size by laser granulometry, major and trace element composition by X-ray fluorescence spectrometry (XRF), bio-assay and more specific studies of Fe, C and S species. In the trench section at this site the top 20 cm of sediment was dark-grey brown in colour with abundant plant roots, grading downwards into slightly mottled, dark grey-brown mud (profile 9211U, Figure 7.5). Below 110 cm the sediment colour changed to dark grey, but no black monosulphide zone or concretions were encountered. Excavation of a number of other trenches and shallow boring investigations during the period 1993–95 have shown that this pattern of sediment zonation is widespread in the Upper Marsh. Areas of black, reduced sediment are typically restricted to the muddy banks and bottoms of small creeks and salt pans which contain standing water for long periods or which are regularly flushed by tidal waters. Such a pocket of soft, black mud occurred within the creek adjacent to the sampled trench section. Samples were also collected from a pit dug in these deposits (profile UCRK, Figure 7.5) for comparative analysis.

Samples for geochemical and microbial analysis were collected using a spatula, sealed in plastic bags, and placed in dry ice for transportation to the laboratory. A 20–30 g sub-sample from each sampling interval was treated *in situ* by adding zinc acetate (10% v/v) to fix easily oxidisable sulphide. After freeze drying and grinding, the samples were split for geochemical, mineralogical and microbial analyses. Figure 7.6 presents a summary of the sequence of procedures used. Organic carbon was determined using a modified Walkley–Black method (Gaudette *et al.*, 1974). Reduced sulphur species (acid-volatile sulphur AVS and pyrite sulphur) were determined on zinc acetate fixed, freeze-dried sediment using a sequential digestion method adapted from Canfield *et al.* (1986). Reactive iron was determined in the Upper Marsh samples using two extractants, sodium dithionite in a buffer of 0.35 M acetic acid and 0.2 M

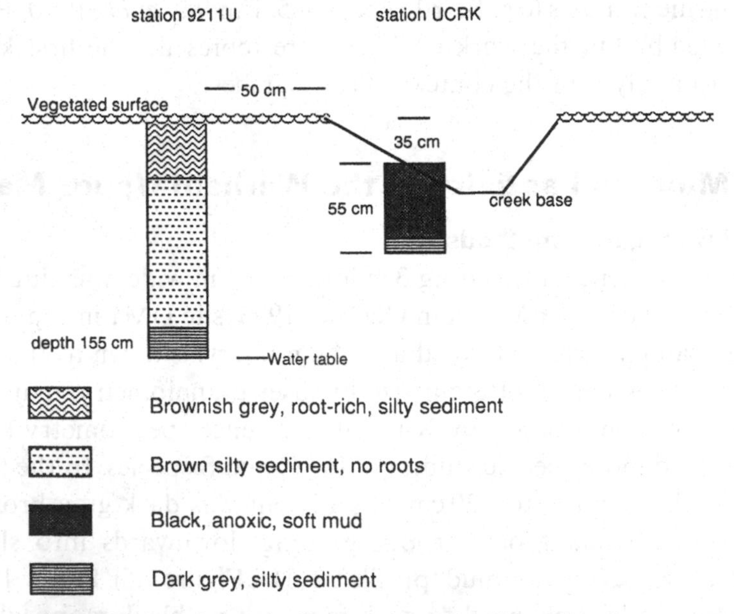

Figure 7.5 Schematic diagram showing the sediment sequence in the Upper Marsh pit (site UM1) and creek bank (UCRK) profiles.

sodium citrate at pH 4.8 (Mehra & Jackson, 1960; Canfield, 1989), and 0.2 M ammonium oxalate and 0.1 M oxalic acid solution at pH 2 (McKeague & Day, 1966).

For the microbial assay, lipids were extracted using a single phase chloroform/methanol/water extraction after Bligh & Dyer (1959) as modified by White *et al.* (1979) to include a phosphate buffer. The total lipid was fractionated into neutral lipid, glyco-lipid and polar lipid fractions by silicic acid column chromatography (SACC) (Parker *et al.*, 1982; White *et al.*, 1983) using the three solvents of increasing polarity (chloroform, acetone and methanol). The polar lipid fraction (PLFA) was

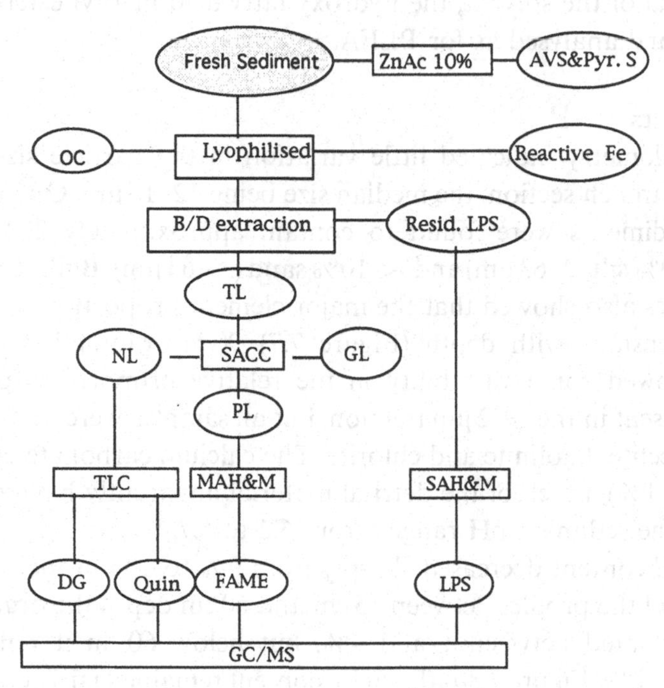

Figure 7.6 Flow diagram summarising the sequence of analytical techniques used.
The abbreviations used in the diagram:
AVS: Acid-volatile sulphides
Pyr. S: pyrite sulphide
OC: organic carbon
Reactive Fe: reactive iron
B/D: Bligh/Dyer extraction
Resid. LPS: residual lipopolysaccharide fatty acids
TL: total lipid
NL: neutral lipid
SACC: silicic acid column chromatography
GL: glycolipid
PL: polar lipid
TLC: thin layer chromatography
MAH&M: mild alkaline hydrolysis and methanolysis
SAH&M: strong acid hydrolysis and methanolysis
DG: digliceride fatty acid
Quin: quinones
FAME: fatty acid methyl esters
LPS: lipopolysaccharide fatty acids
GC/MS: gas chromatography/mass spectroscopy

treated with a mild alkaline methanolysis (MAM) to cleave the glyceride fatty acids and replace their glycerol bond with methyl esters, creating fatty acid methyl esters (FAME). After further processing and separation, the fatty acid methyl esters were quantified by capillary gas chromatography and the peak identifications verified by gas chromatography/mass spectrometry (GC/MS). Diglyceride fatty acids (DGFA) and quinones were recovered from the chloroform-extracted fraction using a thin layer chromatography (TLC) technique for further elutriation and separation as applied to the PLFA. Lippolysaccharide hydroxy fatty acids (LPS OHFA) were released from the lipid-extracted residues by strong acid methanolysis and hexane extraction (Hedrick, Guckert & White, 1991). After removal of the solvent, the hydroxy fatty acid methyl esters were derivatised and analysed as for PLFA.

7.4.2 Results

Laser granulometry indicated little variation in the particle size with depth in the trench section, the median size being 12–15 μm. On average the bulk sediments were found to contain approximately 20% clay (< 2 μm), 70% silt (2–63 μm) and < 10% sand (> 63 μm). Bulk sediment XRF analyses also showed that the major element proportions are also relatively constant with depth (Figure 7.7). X-ray diffraction (XRD) analyses showed some variability in the relative proportions of clay minerals present in the < 2 μm fraction, but all samples were dominated by illite, smectite, kaolinite and chlorite. The calcium carbonate content was low (< 1%), much of the detrital material apparently having been dissolved. The sediment pH ranged from 6.2 to 6.7.

Organic C content decreased sharply from 9% to 1% in the top few centimetres of the profile. Between 15 cm and 60 cm depth the organic C content fluctuated between 1 and 4%, but below 60 cm it remained constant at *c.* 1% (Figure 7.8*a*). Total Fe content remained fairly constant with depth at 4–5%. The dithionite-extractable Fe content, which includes crystalline and amorphous iron oxides (Canfield, 1989), indicates that the sediments contain significant amounts of Fe oxide phases (Figure 7.8*c*). However, in virtually all samples ammonium oxalate-extractable Fe contributes more than half of the dithionite-extractable Fe, indicating that amorphous and poorly crystalline hydrated Fe oxides are dominant. In the lower, greyish part of the sequence the dithionite and oxalate-extractable Fe contents are slightly lower, probably due to iron reduction in the zone of fluctuating groundwater. Concentrations of elemental S, AVS and pyrite S between the surface and 125 cm depth were found to be

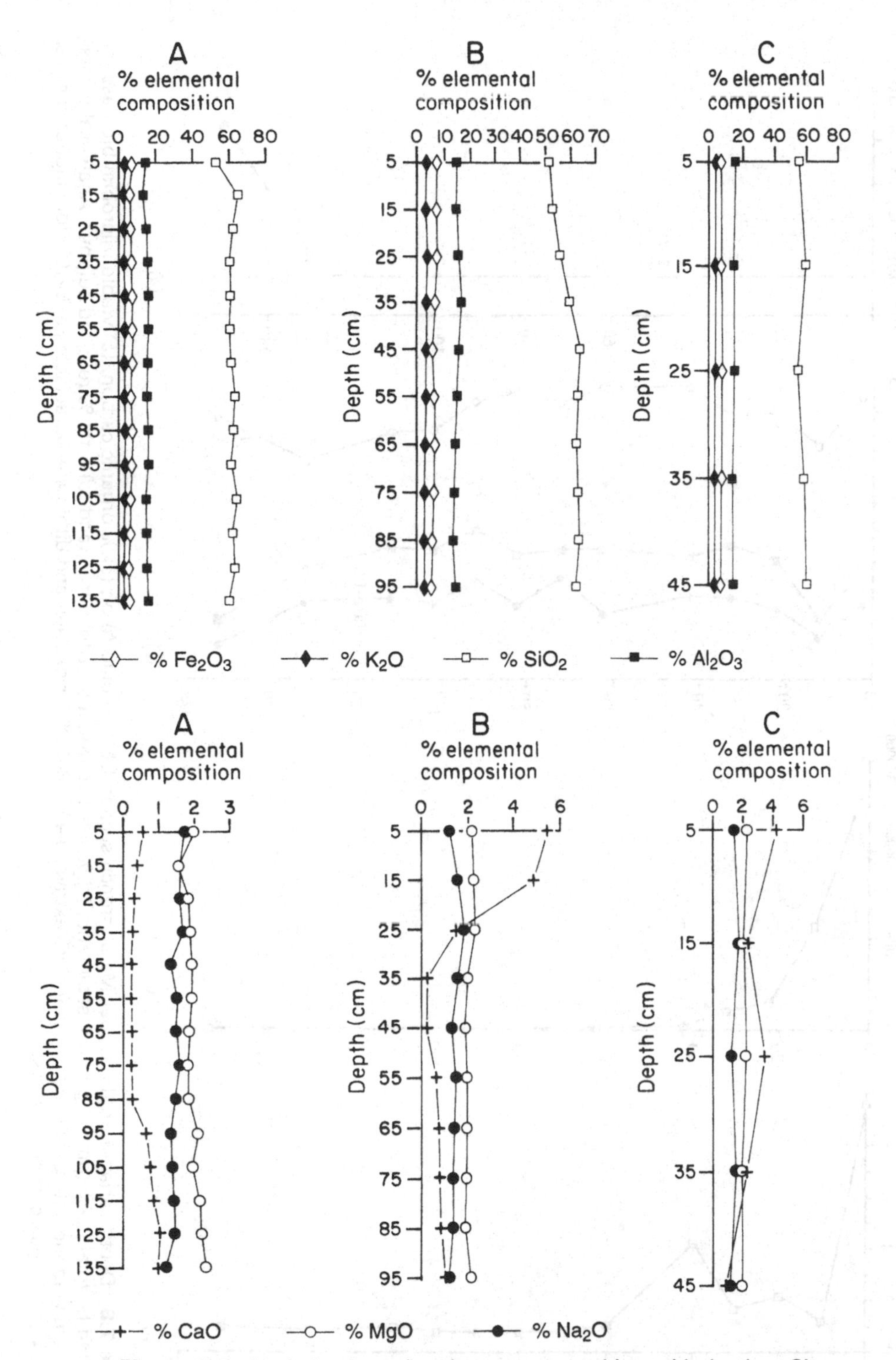

Figure 7.7 Variation in major element composition with depth at Site UM1. Profiles A and B represent the eastern and western ends of the trench section (UM1); Profile C is the nearby creek bank profile (UCRK).

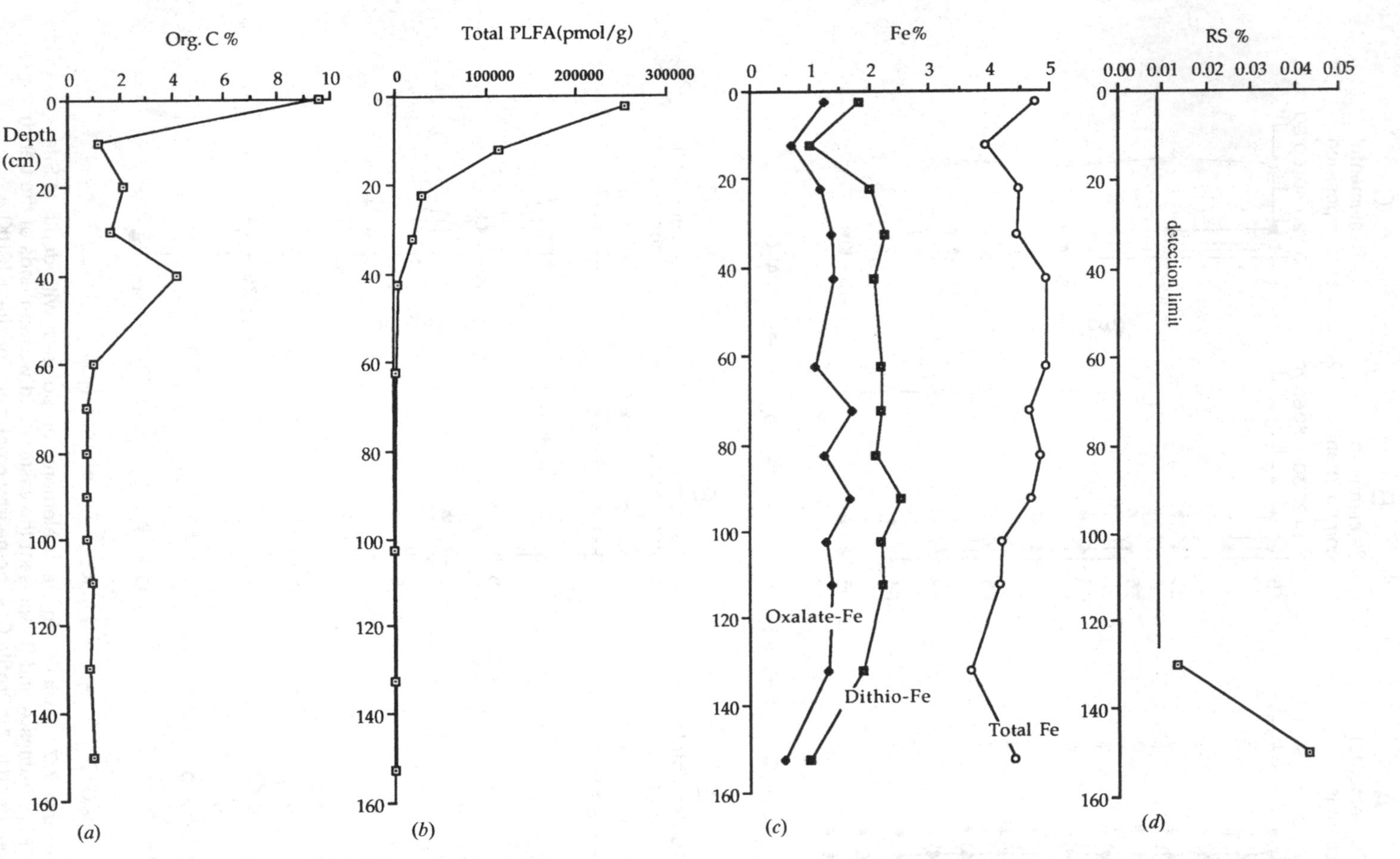

Figure 7.8 Depth variation in the Upper Marsh trench section UM1 (Profile A) of: (*a*) % organic carbon, (*b*) viable microbial biomass indicated by total PLFA (pmol / g of dry sediment), (*c*) % reactive iron and % total iron, and (*d*) total % reduced sulphur. Analytical errors associated with the data points are $< 5\%$ for HCl-extractable Fe, 8–10% for oxalate- and dithionite-extractable Fe, 5–10% for sulphur species, and *c.* 4% for organic carbon values.

below the detection limit (Figure 7.8*d*). Between 125 and 155 cm elemental S and AVS were also undetectable, but total reduced S represented 0.045% of the dry sediment. This suggests that sulphate reduction is a relatively minor process in these sediments at the present time and may never have been very important.

The total PLFA content was found to decrease rapidly with depth in the uppermost 40 cm, showing a markedly steeper decline than organic C (Figure 7.8*b*). This suggests that organic matter becomes less readily metabolisable with depth. Six groups indicative of community structure were identified on the basis of PLFA GC/MS signatures: normal saturates, terminally branched saturates, mid-branched saturates, monoenoics, branched monoenoics, and polyenoics. Normally saturated and unsaturated monoenoic PLFA were found to be dominant. The two groups show inverse abundance with depth, unsaturated monoenoics becoming less abundant down the profile (Figure 7.9*a* & *b*), possibly due to a change in membrane structure from unsaturated to saturated. The distribution of hydroxy fatty acids (OHFA) and the ratio OHFA/PLFA were used to estimate gram negative bacteria contributions to the total microbial community (cf. Parker *et al.*, 1982). The ratio of OHFA/PLFA increased throughout the sediment profile to a point where three times more OHFA were recovered than were PLFA (Figure 7.9*c*). These observations indicate that, although the total number of viable microbial cells decreased with depth, the proportion of gram negative bacterial cells, relative to the total viable biomass, increased with depth. The biomarker for *Desulfobacter* (10Me16:0) shows a peak just below the top of the profile, indicating maximum absolute abundance at this level, and generally declined with depth (Figure 7.9*d*). It shows consistently greater abundance than the biomarker for *Desulfovibrio* (i17:1w7c), which also shows greatest abundance in the upper 30–40 cm (Figure 7.9*e*).

Nutritional stress in bacterial communities is commonly associated with the accumulation of energy-reverse polymers, and a measure of nutritional status can be made by measuring the ratios of lipid storage polymers to cellular biomass. In some bacterial species the ratio of 19 carbon cyclopropyl (cy 19:0) to its monoenoic precursor (18:1w7c) increases with nutrient stress (Tunlid & White, 1992). In the sediment profile studied, this ratio increases rapidly with depth in the upper 40 cm, but falls again in the lower, greyish part of the section (Figure 7.10*a*). A similar trend is shown by other nutritional stress indicators, such as cy17 = 0/16:1w7c, 18:1w7t/18:1w7c and 16:1w7t/16:1w7c (Figures 7.10*b*–*d*). This pattern with depth may reflect the fact that nutrients are

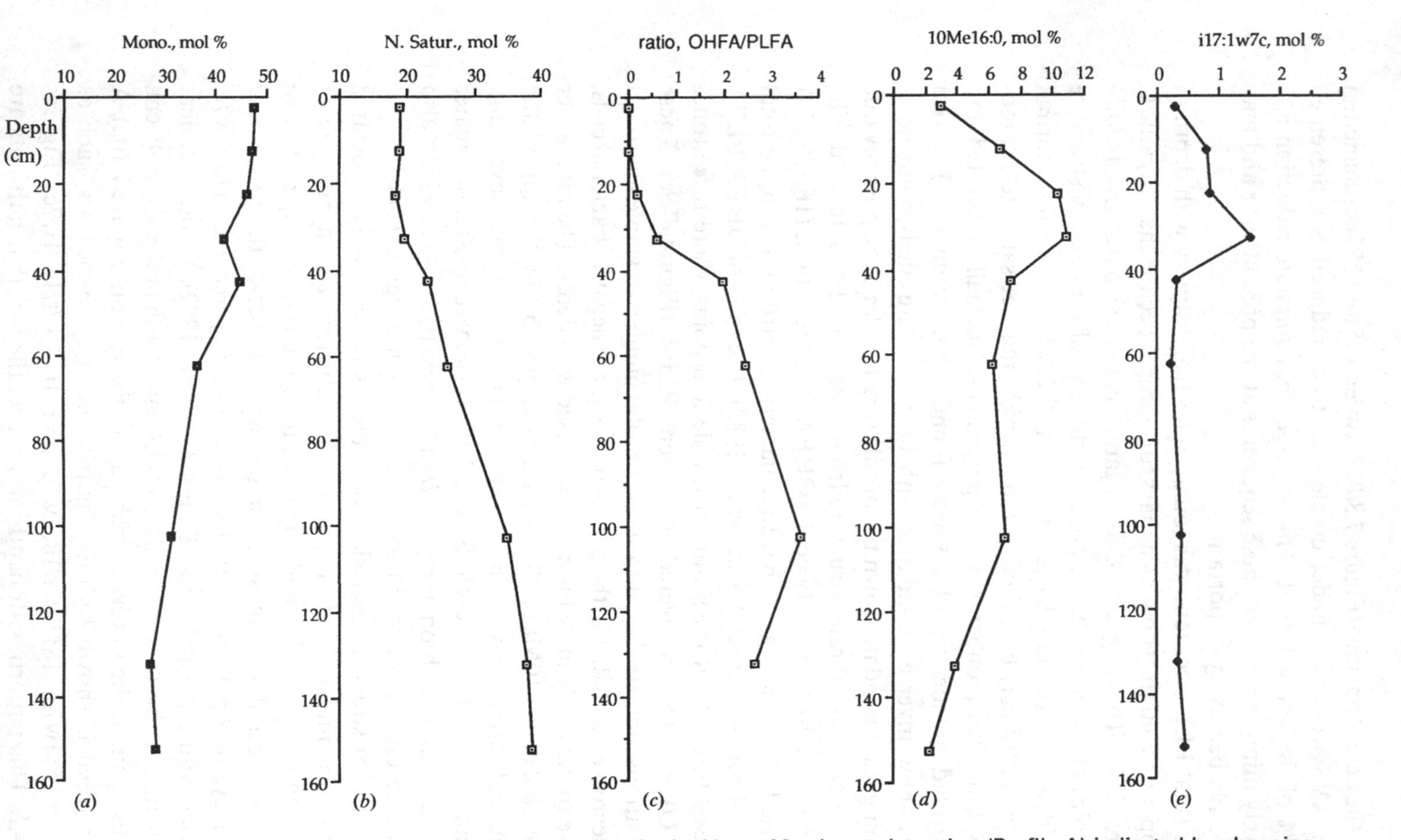

Figure 7.9 Depth variation in microbial community structure in the Upper Marsh trench section (Profile A) indicated by changing abundances of: (*a*) monoenoics, (*b*) normal saturates, (*c*) ratio of OHFA to PLFA, (*d*) 10Me16:0 (biomarker for *Desulfobacter*), and (*e*) i17:1w7c (biomarker for *Desulfovibrio*).

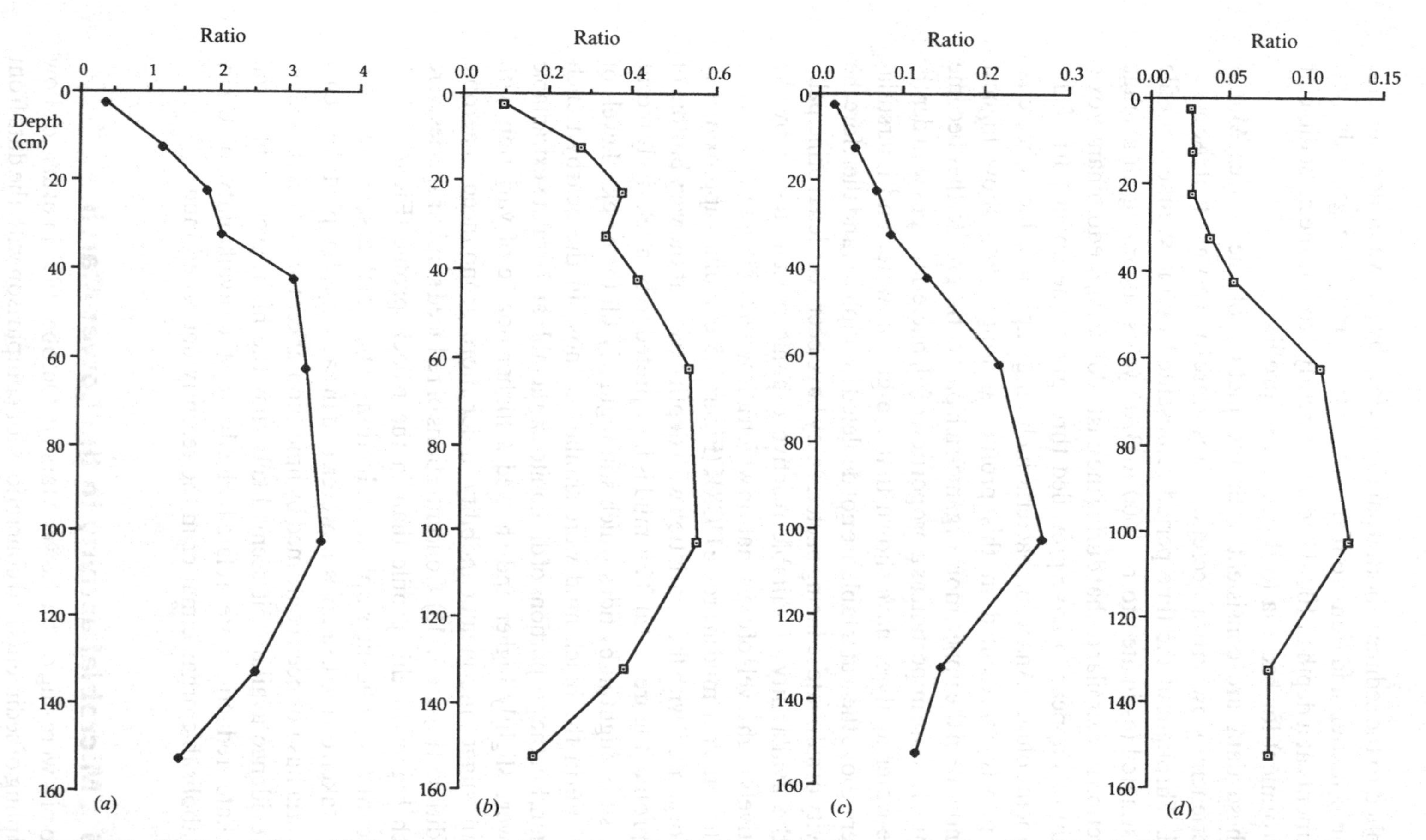

Figure 7.10 Depth variation of indicators of nutritional stress in the Upper Marsh trench section (Profile A, indicated by ratios of: (*a*) cy19:0/18:1wc, (*b*) cy17:0/16:1wc, (*c*) 18:1w7t/18:1w7c and (*d*) 16:1w7t/16:1w7c.

supplied to the sediment both from the surface by sea water flooding and organic matter addition from the vegetation, and from groundwater. Sediments at a depth of 100–110 cm apparently provide the most difficult environment for bacteria to thrive in this profile.

These data are consistent with the picture of the Upper Marsh experiencing slow vertical accretion in response to a very gradual rise in sea level. Throughout the time period represented by the sediment profile investigated (estimated to be 1600 to 3200 years, assuming an average accretion rate similar to that during the past 100 years), predominantly oxic conditions appear to have prevailed throughout the upper part of the sediment column. Anaerobic bacteria, including sulphate reducing bacteria, are poorly represented in this profile; although they show highest abundance in the upper, more organic-rich part of the profile, they become relatively more important as a proportion of all bacterial types with depth. Pore-water analyses have shown that the groundwater and interstitial waters above the water table are not depleted in sulphate, and therefore the activity of sulphate reducing bacteria may have been restricted principally by the availability of suitable reactive organic matter and/or by the relatively high levels of oxygenation which allow aerobic bacteria to thrive.

The soft black mud in profile UCRK (Figure 7.5) contained approximately 2% organic C with little variation with depth except at the very bottom of the profile (Figure 7.11*a*). This mud is interpreted to be a relatively recent deposit brought in by tides which still regularly fill the creeks. Levels of total Fe in the black mud were similar to those in the nearby trench section, but the proportions of dithionite-extractable and oxalate-extractable Fe were slightly higher and showed a higher degree of variation with depth, suggesting greater mobility under more pronounced anaerobic conditions (Figure 7.11*b*). Concentrations of reduced sulphur species were much higher in this profile than in the trench profile (Figure 7.11*c*), indicating significant sulphate reduction. Microbial assays were not undertaken on these samples, however, and the activity of sulphate-reducing bacteria has not been confirmed by lipid biomarker techniques. Based on the evidence available, it seems likely that the high levels of bacterial sulphate reduction are related chiefly to the availability of easily metabolisable organic matter in the recently deposited mud.

7.5 Microbial activity in the Lower Marsh

Two pits were dug on the Lower Marsh in October 1993 in order to allow sampling of sediments and concretions for comparison with the data from

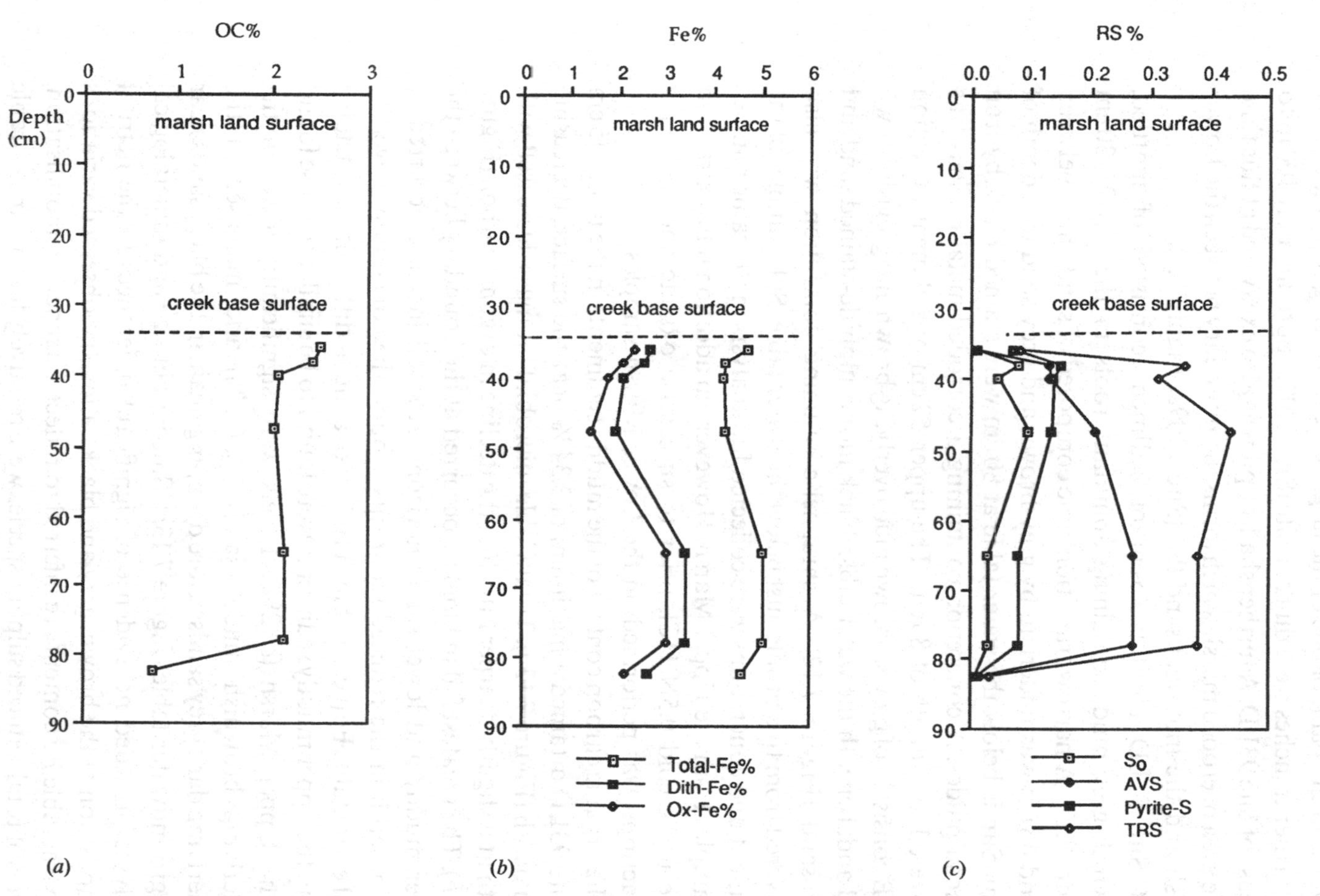

Figure 7.11 Depth variation within the Upper Marsh Creek section (Profile UCRK): (*a*) % organic carbon; (*b*) % reactive iron and total iron, and (*c*) % reduced sulphur species.

the Upper Marsh. Site 93LD was located in the mid part of the Lower Marsh to the east of the Cocklestrand footpath (Figure 7.1), approximately 15 m away from the nearest creek. Site 93CB was located 100 m further seaward on the bank of one of the larger creeks. During September 1994 two further trenches were dug at a site (Site 9409) approximately 100 m to the east of Site 93LD. A further shallow pit was also excavated at this time in the area of eroded marsh which forms the boundary between the Lower Marsh and the intertidal sand flat (Site 9409a, Figure 7.1).

At Site 93LD the top 45 cm of sediment consisted of mottled, brownish-grey mud, containing abundant roots in the top 25–30 cm (Figure 7.12). A narrow transition zone composed of sandy mud between 45 and 48 cm was underlain by grey-brown and grey sand to a depth of 80 cm. Sands below the water table at 80 cm were stained black by iron monosulphides. A concretion containing a corroded metal nucleus was recovered at a depth of 85 cm. The upper 52 cm of sediment in section 93CB consisted of greyish-brown mud overlying brown and greyish-brown sand and, beneath the water table, black monosulphide-stained sand and grey sand (Figure 7.12). A shrapnel-containing concretion was also recovered from this profile, just below the water table. Sub-samples of the sediment and concretions were collected for analysis in the same manner as samples from the Upper Marsh. However, an additional Fe extraction stage using cold 0.5N HCl, prior to successive oxalate and dithionite extractions, was performed on the Lower Marsh samples.

The organic carbon content of the muddy sediments in the upper 45 cm at Site 93LD attained a maximum of 3.34% near the surface, decreasing with depth (Figure 7.13*a*). Total viable microbial biomass decreased with depth at a slightly sharper rate than the decline in organic carbon (Figure 7.13*b*). The greatest fall in biomass occurred at the boundary between the upper muddy and lower sandy horizons. A small increase occurred at 85 cm depth in the reduced sediments around the concretion. Levels of total extractable Fe (i.e. combined HCl + oxalate + dithionite-extractable Fe) in the upper muddy sediments were found to be similar to those found in the Upper Marsh (*c.* 2%). Levels were significantly lower in the underlying brownish muddy sands ($< 0.5\%$) and increased in the underlying dark grey sands before dropping again in the black sands near the groundwater table (Figure 7.13*c*). Since grain size and mineralogical analyses provided no evidence of significant differences in the detrital composition of the brown, grey and black sands, the observed variations in extractable Fe content are inferred to reflect diagenetic remobilisation. Levels of total reduced sulphur species were relatively low but measurable

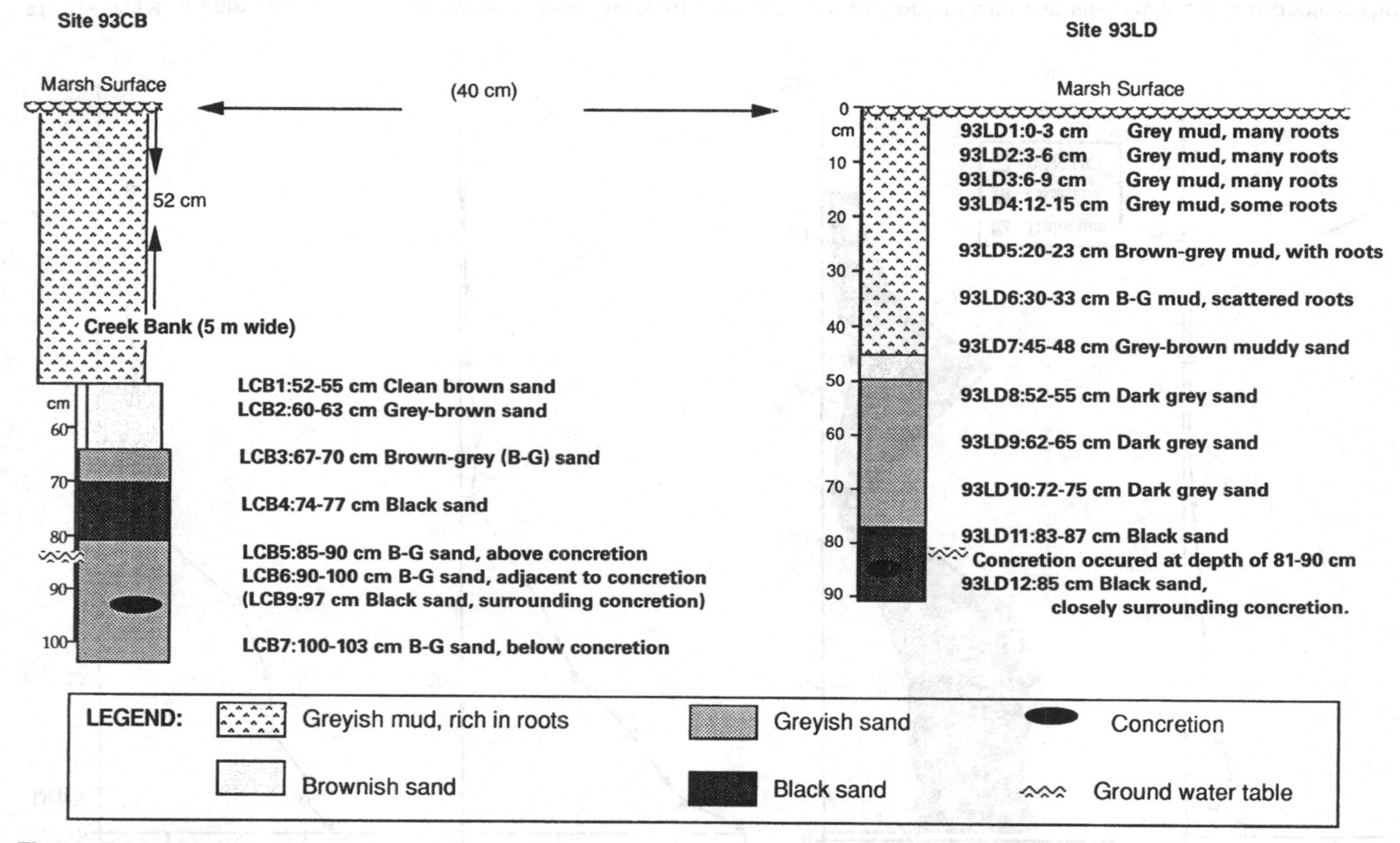
Site 93CB
Marsh Surface
(40 cm)
52 cm
Creek Bank (5 m wide)
cm
60
70
80
90
100
LCB1:52-55 cm Clean brown sand
LCB2:60-63 cm Grey-brown sand
LCB3:67-70 cm Brown-grey (B-G) sand
LCB4:74-77 cm Black sand
LCB5:85-90 cm B-G sand, above concretion
LCB6:90-100 cm B-G sand, adjacent to concretion
(LCB9:97 cm Black sand, surrounding concretion)
LCB7:100-103 cm B-G sand, below concretion
Site 93LD
Marsh Surface
0
cm
10
20
30
40
50
60
70
80
90
93LD1:0-3 cm Grey mud, many roots
93LD2:3-6 cm Grey mud, many roots
93LD3:6-9 cm Grey mud, many roots
93LD4:12-15 cm Grey mud, some roots
93LD5:20-23 cm Brown-grey mud, with roots
93LD6:30-33 cm B-G mud, scattered roots
93LD7:45-48 cm Grey-brown muddy sand
93LD8:52-55 cm Dark grey sand
93LD9:62-65 cm Dark grey sand
93LD10:72-75 cm Dark grey sand
93LD11:83-87 cm Black sand
Concretion occured at depth of 81-90 cm
93LD12:85 cm Black sand,
closely surrounding concretion.
LEGEND:
Greyish mud, rich in roots
Greyish sand
Concretion
Brownish sand
Black sand
Ground water table

Figure 7.12 Diagram showing the sediment profiles at Site 93CB and Site 93LD on the Lower Marsh.

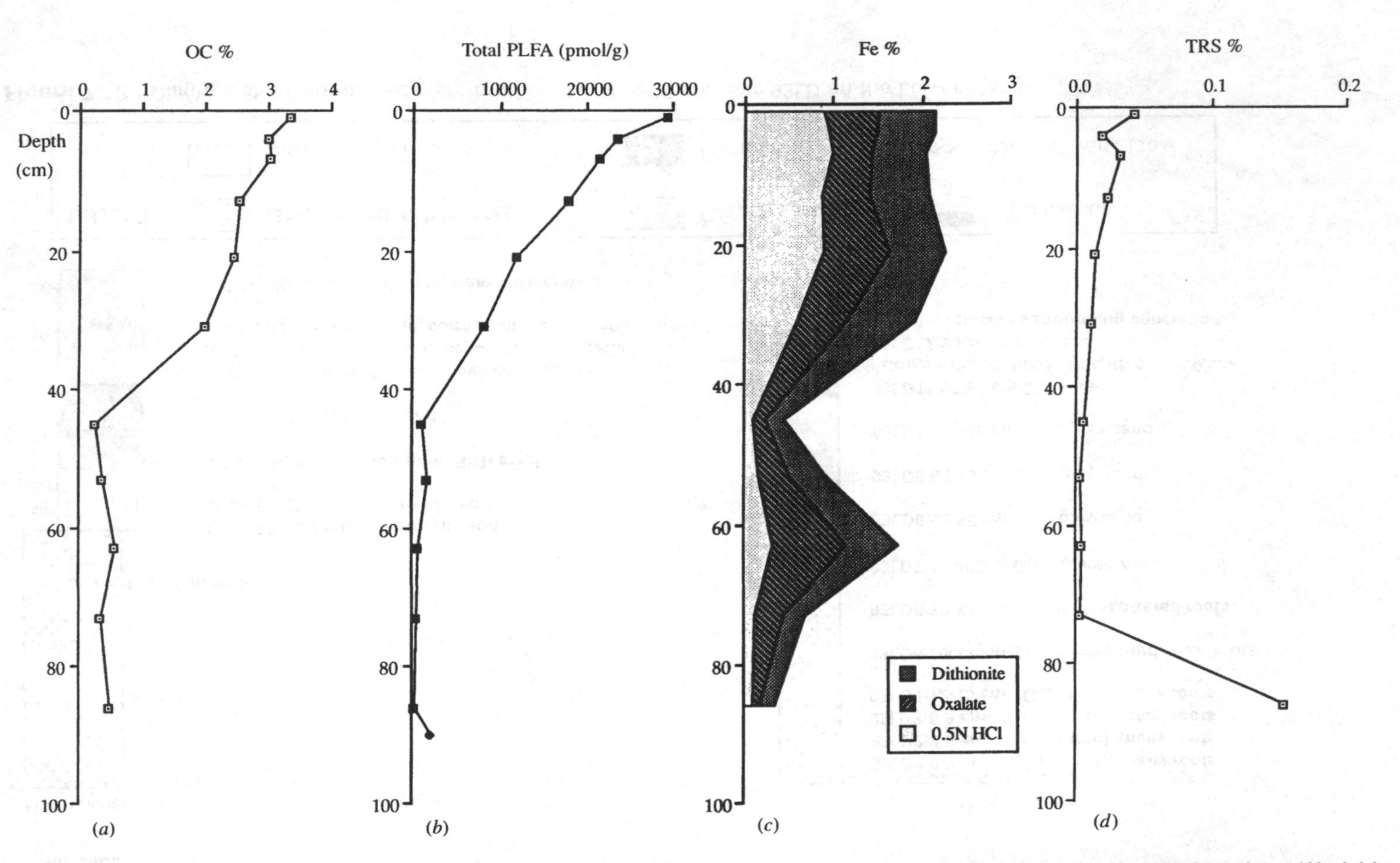

Figure 7.13 Depth variation in geochemical parameters and indicators of microbial activity at site 93LD: (*a*) % organic carbon; (*b*) viable microbial biomass, indicated by total PLFA, (*c*) % total reactive iron determined by sequential extraction, (*d*) % total reduced sulphur.

in the upper part of the profile, showing a progressive reduction with depth and a sudden increase in the black sands adjacent to the concretion (Figure 7.13*d*).

The bioassay results showed that, in general, the proportion of terminally branched saturates increased slightly, and that of monoenoics decreased, with depth in the profile (Figure 7.14*a* and *b*). The OHFA/PLFA ratio increased with depth in the muddy part of the profile, suggesting more significant gram negative populations (Figure 7.14*c*). Levels of OHFA fell again in the upper, brownish-grey, sandy sediments and showed a small secondary peak in the zone of black sands. The biomarkers for *Desulfobacter* (10Me16:0) and *Desulfovibrio* (i17:1w7c) showed the former in greater relative abundance throughout the profile (Figure 7.14*d* and *e*), as was observed on the Upper Marsh. *Desulfobacter* are apparently relatively most abundant in the upper muddy sediments and the grey sands, while *Desulfovibrio* become relatively more significant in the black sands around the concretion. An increase in the ratio of cy19:0/18:1w7c with increasing depth was also observed, suggesting an increase in nutritional stress of the microbial community (Figure 7.14*f*). This may indicate the presence of an aged microbial population and/or an increase in anaerobic respiration. A slight reduction in apparent stress is indicated in the mid part of profile 93LD, suggesting increased availability of nutrients and/or oxygen levels within the uppermost sandy sediments. Stress levels then increase again in the underlying grey sands which are low in reactive Fe and organic C, before decreasing again in the black, reactive Fe and reduced sulphur-rich sands close to the water table.

In profile 93CB, organic carbon content increased dramatically with depth below the surface of the creek bank (Figure 7.15*a*). Total microbial biomass showed a reverse trend, however (Figure 7.15*b*). Levels of total extractable Fe were lower than in profile 93LD and were relatively constant with depth (Figure 7.15*c*). The abundance of total reduced sulphur species increased dramatically with depth, attaining a maximum value in the black sands surrounding the concretion (Figure 7.15*d*). Indicators of community structure show a slight decrease in the abundance of monoenoics and an increase in terminally branched saturates with depth (Figure 7.16*a* and *b*), as also seen in profile 93LD. The OHFA/PLFA ratio showed a slight increase in the lower part of the black sand zone (Figure 7.16*c*), suggesting relatively more important gram negative populations. The biomarker for *Desulfobacter* showed a progressive increase in abundance with depth, but that for *Desulfovibrio* peaked in the black sands above the concretion (Figure 7.16*d* and *e*).

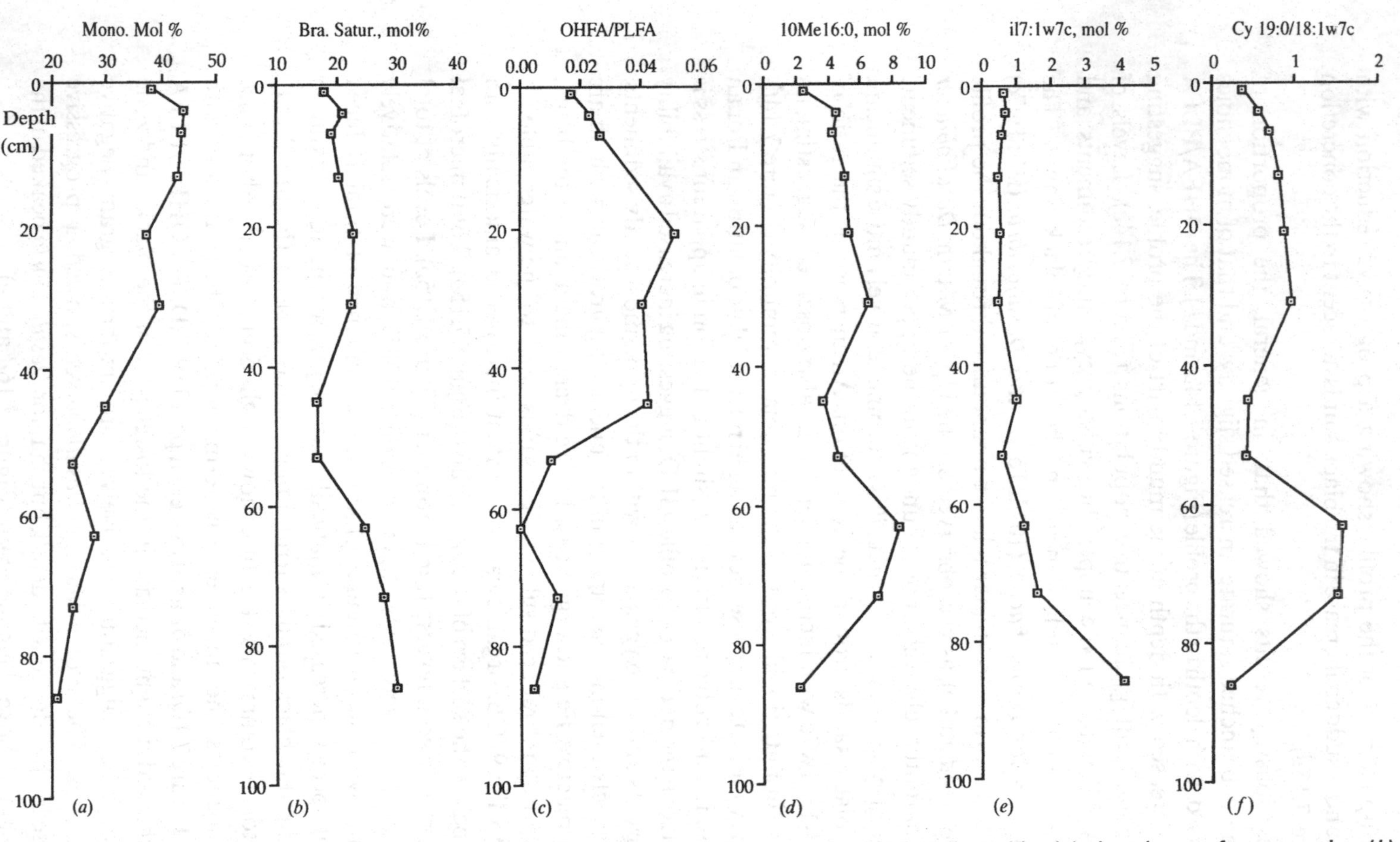

Figure 7.14 Indicators of community structure and environmental stress in profile 93LD indicated by (*a*) abundance of monoenoics, (*b*) abundance of branched saturates, (*c*) the ratio of OHFA to PLFA, (*d*) the biomarker for *Desulfobacter* (10Me16: 0), (*e*) the biomarker for *Desulfovibrio* (i17: 1w7c), and (*f*) the ratio of cy19: 0/18: 1w7c.

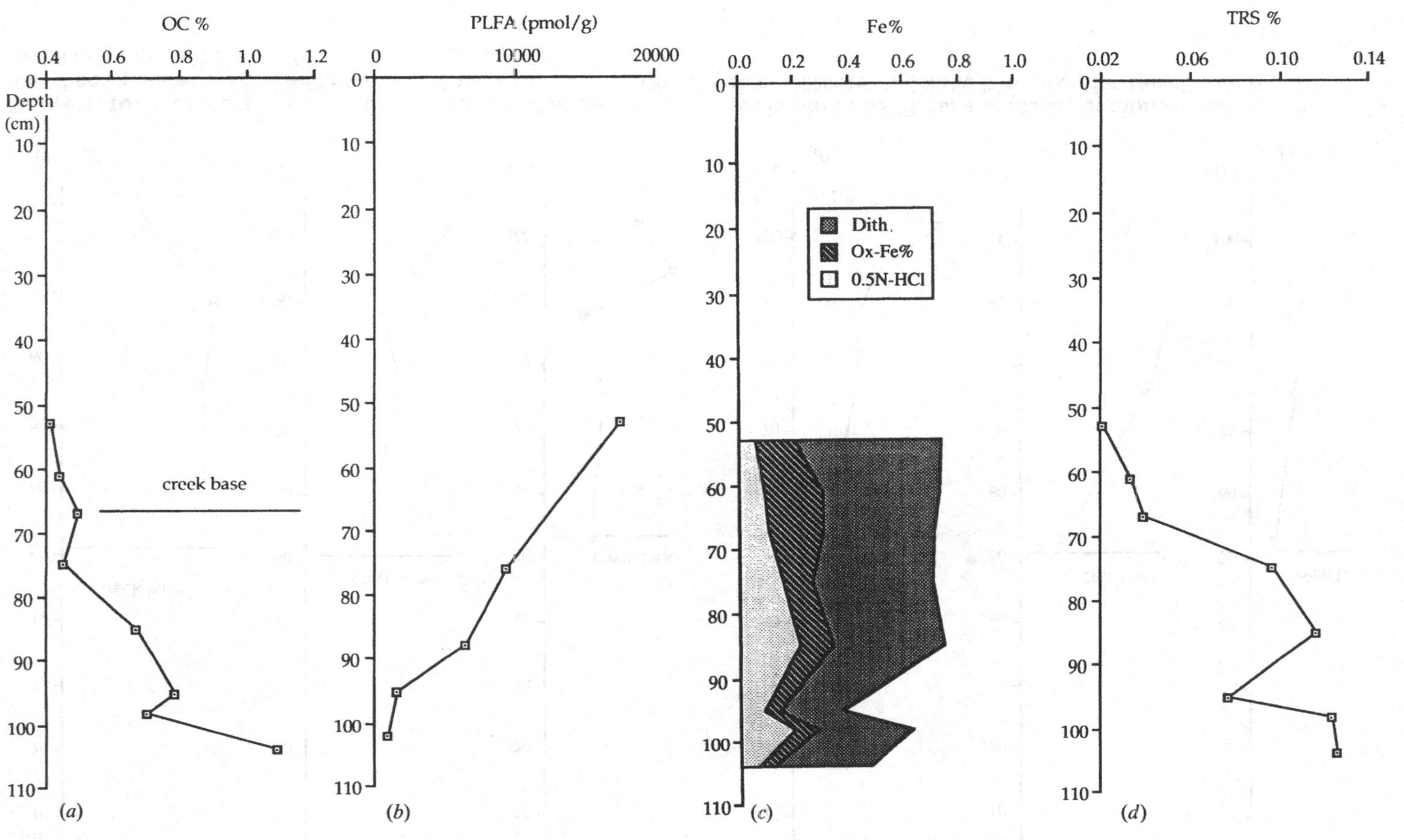

Figure 7.15 Depth variation in geochemical parameters and indicators of microbial activity in profile 93CB: (*a*) % organic carbon, (*b*) viable microbial biomass, indicated by total PLFA, (*c*) % reactive Fe, and (*d*) % total reduced sulphur.

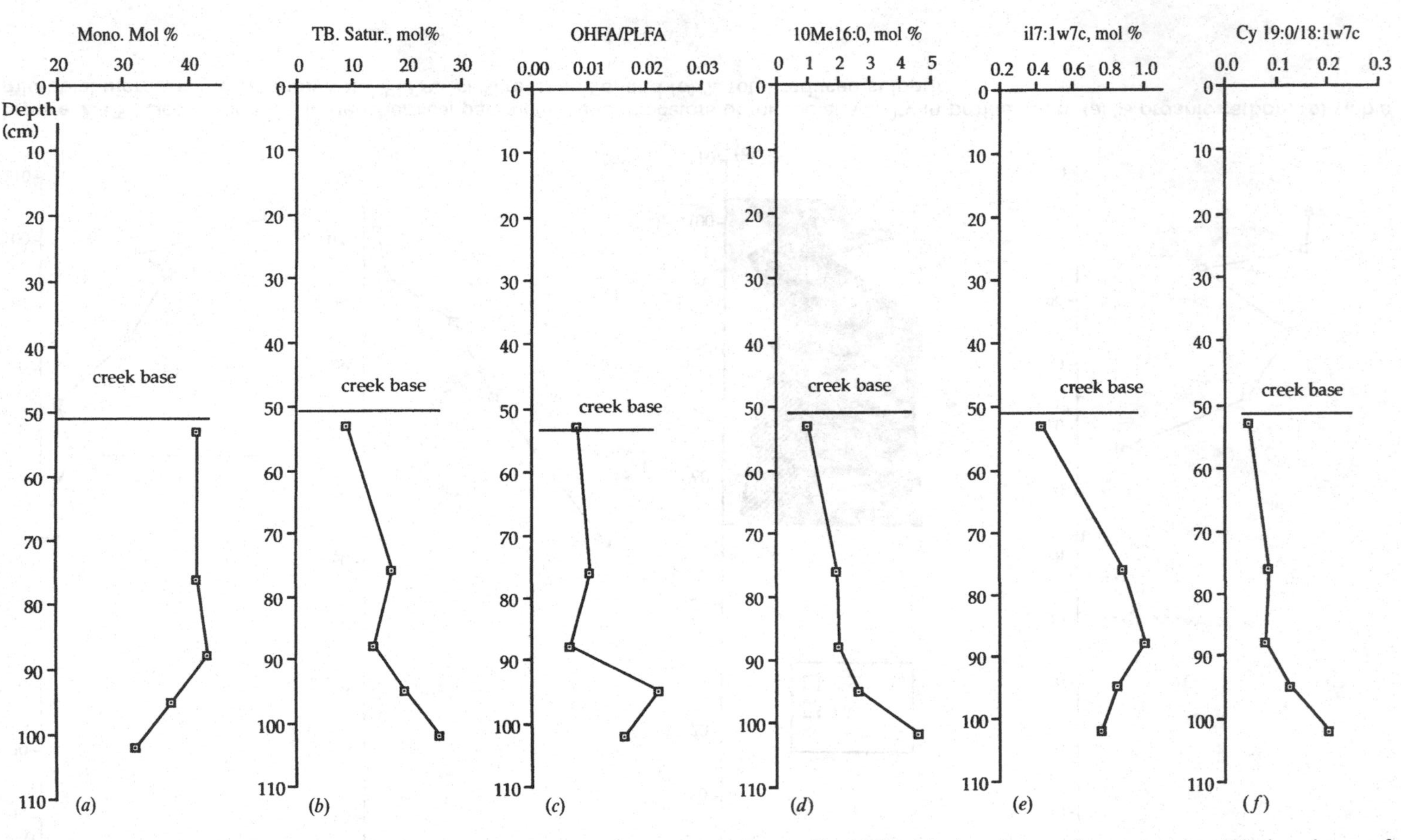

Figure 7.16 Indicators of community structure and environmental stress in profile 93CB: (*a*) abundance of monoenoics, (*b*) abundance of branched saturates, (*c*) OHFA/PLFA ratio, (*d*) the biomarker for *Desulfobacter* (10Me16:0), (*e*) the biomarker for *Desulfovibrio* (i17:1w7c), and (*f*) the ratio of cy19:0/18:lw7c.

These results suggest that anaerobic microbial activity, including sulphate reduction, typically increases with depth in the Lower Marsh muddy sediments but remains of relatively limited importance compared with aerobic microbial processes. Anaerobic activity decreases in the relatively well-drained brown and greyish-brown sands which represent the former intertidal flat, but increases again in the black and grey sands which lie below and immediately above the water table. In the upper muddy sediments the vertical variation in degree of anaerobic activity is probably controlled by the balance between organic carbon content and oxygen availability, both of which decline with depth. In the underlying sandy sediments sea water saturation maintains anaerobic conditions below the water table, but rates of anaerobic microbial decomposition are limited mainly by the availability of suitable organic carbon sources.

In the creek bank sediments, profile 93CB, levels of organic carbon are much lower than in the saltmarsh near-surface sediments, profile 93LD, but levels of reduced sulphur species are significantly higher while total viable microbial biomass is of the same approximate magnitude. This suggests that the form in which the organic carbon is present is of major significance in controlling the rate and nature of microbial activity. Algae, diatoms and other marine species which live in the near-surface creek sediments, and which are regularly introduced to the system by tidal flushing, are probably of great significance in this respect. Analyses of pore-water samples have indicated that levels of sulphate are high throughout the Warham marsh system, and sulphate availability is therefore unlikely to be a limiting factor in sulphate reduction. However, there is clearly a greater opportunity for maintenance and replenishment of the dissolved sulphate source within the near-surface creek-bank sediments and below the saline groundwater table.

In comparing the biomarker signatures of the Upper Marsh and Lower Marsh, the most distinguishing feature is the different rate of decline of total viable microbial biomass with depth in the upper part of the profiles. In the Upper Marsh an order of magnitude decrease in the viable biomass occurred in the upper 20 cm while in the Lower Marsh profile a similar order of magnitude decrease occurred over a vertical distance of *c.* 40 cm. This difference reflects the varying rates of sediment accretion on the two marshes.

Analysis of samples taken from different parts of the carbonate concretions, and from the adjacent host sediment at site 93LD, showed that organic carbon content is higher within the concretions than in the surrounding sediments (Figure 7.17*a*). Little difference was found between

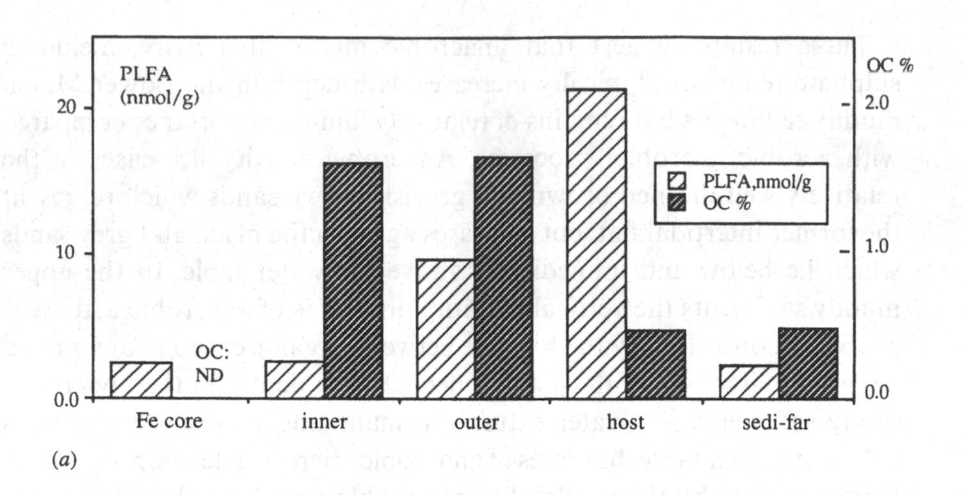

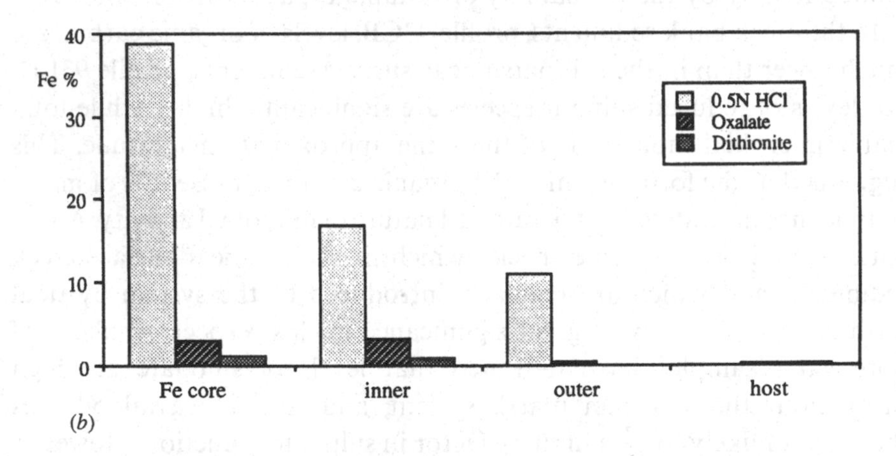

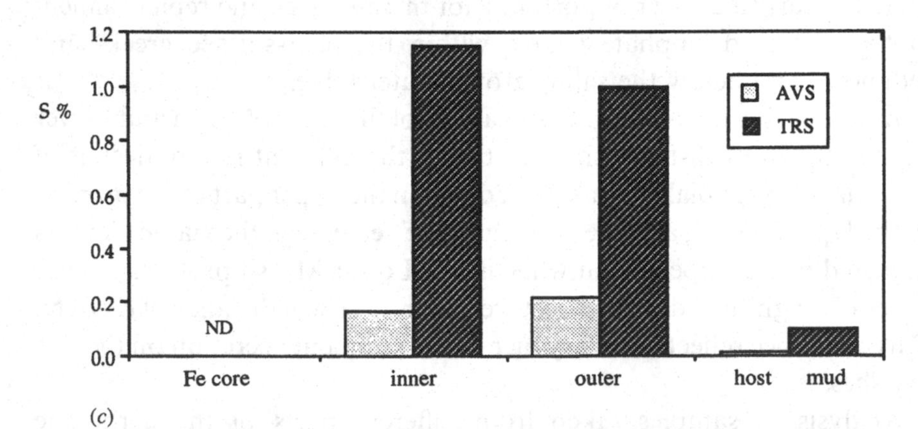

Figure 7.17 Variation in geochemical parameters and indicators of microbial activity within different parts of the concretion and host sediment from site 93LD: (*a*) % organic carbon and viable microbial biomass, (*b*) different forms of extractable iron, (*c*) % acid volatile sulphide (AVS) and % total reduced sulphur (TRS).

the inner and outer portions of the concretion collected from profile 93CB, but no organic carbon was determined in the iron-rich core. Microbial biomass was found to attain a maximum in the sediment immediately surrounding the concretion, but was also high in the outer part of the concretion. Levels of extractable iron were found to decline systematically from the core of the concretion towards the rim (Figure 7.17*b*). Total reduced sulphur levels were found to be high in both inner and outer parts of the concretion compared with the host sediment (Figure 7.17*c*). In terms of community structure, monoenoics were again found to be the dominant group in all parts of the concretion and in the surrounding sediment, with the relative abundance of this group achieving a maximum in the outer part of the concretion (Figure 7.18*a*). The indicators of nutritional stress showed maximum stress conditions in the inner part of the concretion, decreasing to a minimum in the surrounding sediment (Figure 7.18*b*). The biomarkers for *Desulfobacter* and *Desulfovibrio* suggested that the former are relatively more abundant within the concretion and in the sediment immediately adjacent to it, whereas the latter are more abundant in the sediment some distance away at the same stratigraphic level (Figure 7.18*c*). Within the concretion the relative abundance of *Desulfobacter* was found to increase from the core towards the rim, whereas that of *Desulfovibrio* declined. The observed dominance of *Desulfovibrio* in the iron-rich core of the concretion is consistent with the finding that these bacteria are capable of reducing ferrous iron when sulphate supply is limited (Coleman *et al.*, 1993).

7.6 Conclusions

Lipid biomarker techniques have proved to be a powerful tool which, in combination with more conventional geochemical and mineralogical analyses, has shed new light on the diagenetic reactions taking place within the north Norfolk saltmarshes. The results obtained from the most recent sampling on the Lower Marsh and intertidal flats have still to be fully interpreted, but it is clear that there are major differences between the diagenetic regimes of the Upper Marsh and the Lower Marsh which reflect their contrasting accretion histories and sedimentological characteristics.

Acknowledgements

We thank Nigel Pontee, Stephen Crooks and Tim Steel of Reading University for assistance with field sampling, and David White, David

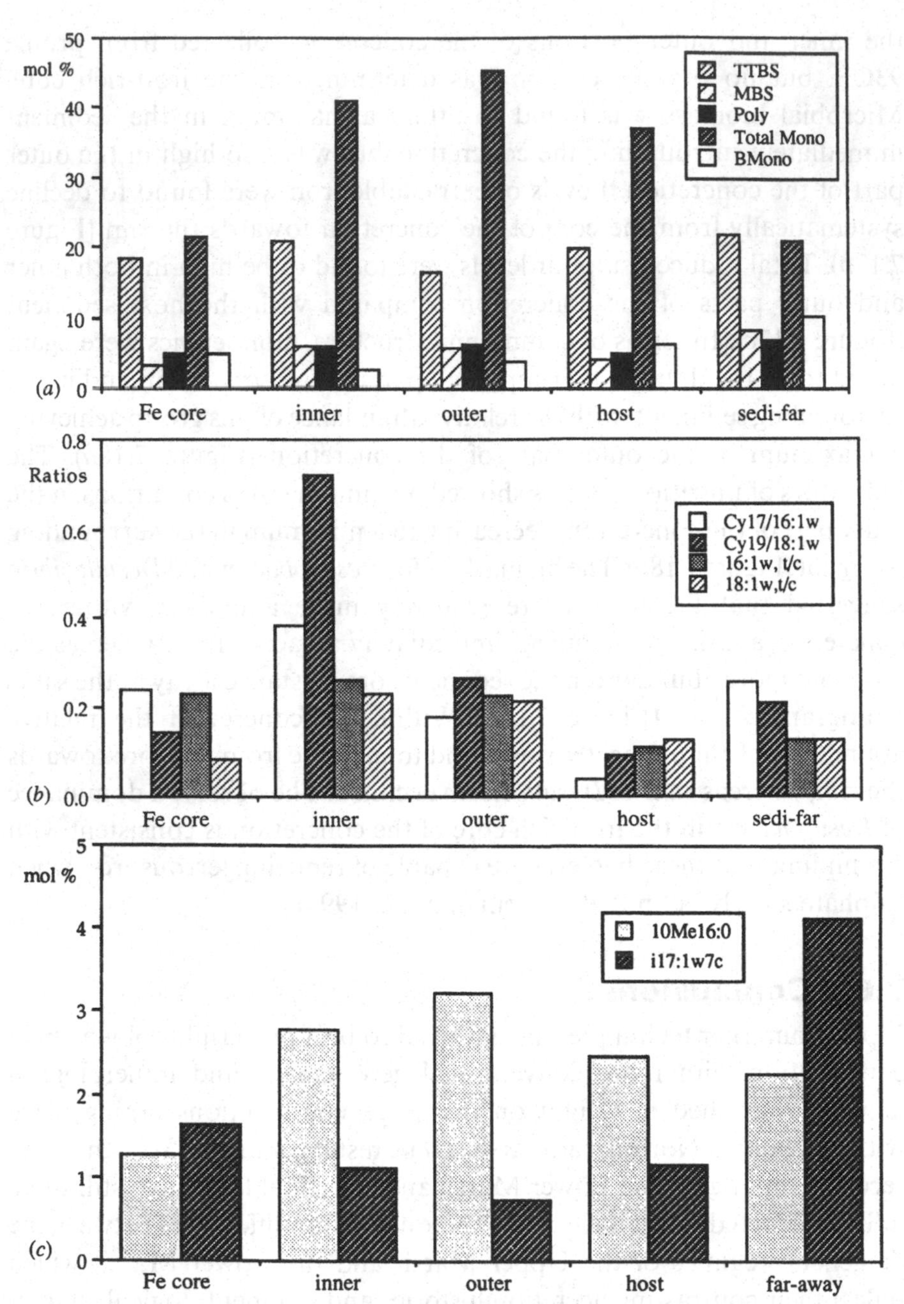

Figure 7.18 Variation in microbial community structure and indicators of environmental stress within different parts of the concretion and host sediment from profile 93LD: (*a*) community structure –TBS = terminally branched saturates, MBS = mid-branched saturates; Poly = polyenoics, Total Mono = total monoenoics, BMono = branched monoenoics. Sedi-far represents the host sediment at some distance from the concretion but at the same stratigraphic level; (*b*) indicators of nutritional stress; (*c*) biomarkers for *Desulfobacter* (10Me16: 0) and *Desulfovibrio* (i17: 1w7c).

Hedrick and David Ringelberg of CEB, Knoxville, Tennessee, for assistance with the bioassays. Fieldwork on the Warham marshes is undertaken by permission of English Nature. The work was supported by NERC Grants GR3/5476 and GR3/5476A. This paper represents University of Reading PRIS Contribution No. 414.

References

Al-Agha, M.R., Burley, S.D., Curtis, C.D. & Esson, J. (1995) Complex cementation textures and authigenic mineral assemblages in recent concretions from the Lincolnshire Wash (east coast, UK) driven by Fe(0) to Fe(II) oxidation. *Journal of the Geological Society, London*, **152**, 157–71.

Allison, P.A. & Pye, K. (1994) Early diagenetic mineralization and fossil preservation in modern carbonate concretions. *Palaios* **9**, 561–75.

Andrews, J.E. (1988) Methane-related Mg–calcite cement in recent tidal flat sediments from the Firth of Forth. *Scottish Journal of Geology* **24**, 233–44.

Baird, B.H., Nivens, D.E., Parker, J.H. & White, D.C. (1985) The biomass, community structure, and spatial distribution of the sedimentary macrobiota from a high energy area of the deep sea. *Deep Sea Research* **32**, 1089–99.

Berner, R.A. & Westrich, J.T. (1985) Bioturbation and the early diagenesis of carbon and sulfur. *American Journal of Science* **285**, 193–206.

Blackburn, T.H. & Blackburn, N.D. (1993) Rates of microbial processes in sediments. *Philosophical Transactions of the Royal Society of London* A, **344**, 49–58.

Bligh, E.G. & Dyer, W.J. (1959) A rapid method of total lipid extraction and purification. *Canadian Journal of Biochemistry and Physiology* **37**, 911–17.

Canfield, D.E. (1989) Reactive iron in marine sediments. *Geochemica Cosmochimica Acta* **53**, 619–32.

Canfield, D.E., Raiswell, R., Westrich, J.T., Reaves, C.M. & Berner, R.A. (1986) The use of chromium reduction in the analysis of reduced inorganic sulfur in sediments and shales. *Chemical Geology* **16**, 59–62.

Coleman, M.L., Hedrick, D.B., Lovley, D.R., White, D.C. & Pye, K. (1993) Reduction of Fe(III) in sediments by sulphate reducing bacteria. *Nature* **361**, 436–8.

Davison, W., Grimes, G.W., Morgan, J.A.W. & Clarke, K. (1991) Distribution of dissolved iron in sediment pore waters at submillimetre resolution. *Nature* **252**, 323–5.

Duck, R.W. (1995) Subaqueous shrinkage cracks and early sediment fabrics preserved in Pleistocene calcareous concretions. *Journal of the Geological Society, London*, **152**, 151–6.

Eglinton, G., Parkes, R.J. & Zhao, M. (1993) Lipid biomarkers in biogeochemistry: future roles? *Marine Geology* **113**, 141–5.

Findlay, R.H., Trexler, M.B., Guckert, J.B. & White, D.C. (1990) Laboratory study of disturbance in marine sediments: response of a microbial community. *Marine Ecological Progress Series* **62**, 121–33.

Funnell, B.M. & Pearson, I. (1989) Holocene sedimentation on the north Norfolk barrier coast in relation to relative sea level change. *Journal of Quaternary Science* **4**, 25–36.

Gaudette, H.E., Flight, W.R., Toner, L. & Folger, D. (1974) An inexpensive titration method for the determination of organic carbon in recent sediments. *Journal of Sedimentary Petrology* **44**, 249–53.

Guckert, J.B., Antworth, C.P., Nichols, P.D. & White, D.C. (1985) Phospholipid, ester-linked fatty acid profiles as reproducible assay for changes in prokaryotic community structure of estuarine sediments. *FEMS Microbiology Ecology* **31**, 147–58.

Hedrick, D.B., Guckert, J.B. & White, D.C. (1991) Archaebacterial ether lipid diversity: analysis by supercritical fluid chromatography. *Journal of Lipid Research* **32**, 656–66.

Jorgensen, B.B. (1978) A comparison of methods for the quantification of bacterial sulfate reduction in coastal marine sediment. II. Estimation from chemical and bacteriological field data. *Geomicrobiology* **1**, 45–64.

Jorgensen, B.B. & Bak, F. (1991) Pathways and microbiology of thiosulfate transformations and sulfate reduction in a marine sediment (Kattegat, Denmark). *Applied and Environmental Microbiology* **57**, 847–56.

Lovley, D.R. (1991) Dissimilatory Fe(III) and Mn(IV) reduction. *Microbiology Reviews* **55**, 259–87.

McKeague, J.A. & Day, J.H. (1966) Dithionite and oxalate extractable Fe and Al as aids in differentiating various classes of soils. *Canadian Journal of Soil Science* **46**, 13–22.

Mehra, O.P. & Jackson, M.L. (1960) Iron oxide removal from soils and clays by a dithionite–citrate system buffered with sodium bicarbonate. *Clays and Clay Minerals* **7**, 317–27.

Parker, J.H., Fredrickson, H.L., Vestal, J.R. & White, D.C. (1982) Sensitive assay, based on hydroxy-fatty acids lipopolysaccharide lipid A for gram negative bacteria in sediments. *Applied and Environmental Microbiology* **44**, 1170–7.

Parkes, R.J., Cragg, B.A., Getliff, J.M., Harvey, S.M., Fry, J.C., Lewis, C.A. & Rowland, S.J. (1993a) A quantitative study of microbial decomposition of biopolymers in recent sediments from the Peru Margin. *Marine Geology* **113**, 55–66.

Parkes, R.J., Dowling, N.J.E., White, D.C., Herbert, R.A. & Gibson, G.R. (1993b) Characterization of sulphate-reducing bacterial populations within marine and estuarine sediments with different rates of sulphate reduction. *FEMS Microbiology Ecology* **102**, 235–50.

Purchas, A.W. (1965) *Some History of Wells-next-the-Sea and District*. East Anglian Magazine Ltd, Ipswich, 140 pp.

Pye, K. (1981) Marshrock formed by iron sulphide and siderite cementation. *Nature* **294**, 650–2.

Pye, K. (1984) SEM analysis of siderite cements in intertidal marsh sediments, Norfolk, England. *Marine Geology* **56**, 1–12.

Pye, K. (1988) An occurrence akaganeite (β-FeO–OH.Cl) in recent oxidized carbonate concretions, Norfolk, England, *Mineralogical Magazine* **52**, 125–6.

Pye, K. (1992) Saltmarshes on the barrier coastline of north Norfolk, eastern England. In *Saltmarshes: Morphodynamics, Conservation and Engineering Significance*, ed. J.R.L. Allen & K. Pye. Cambridge University Press, Cambridge, pp. 148–78.

Pye, K., Dickson, J.D., Schiavon, N., Coleman, M.L. & Cox, M. (1990) Formation of siderite–Mg calcite–iron sulphide concretions in intertidal marsh and sandflat sediments, north Norfolk, England. *Sedimentology* **37**, 325–43.

Sorenson, J. (1987) Nitrate reduction in marine sediment: pathways and interaction with iron and sulfur cycling. *Geomicrobiology Journal* **5**, 401–21.

Tunlid, A. & White, D.C. (1992) Biochemical analysis of biomass, community structure, nutritional status, and metabolic activity of the microbial community in soil. In *Soil Biochemistry*, Volume 7, eds. G. Stotzky & J.M. Bollag, Marcel Dekker, New York, pp. 229–62.

Vernberg, F.J. (1993) Salt-marsh processes: a review. *Environmental Toxicology and Chemistry* **12**, 2167–95.

White, D.C. (1993) In situ measurement of microbial biomass, community structure and nutritional status. *Philosophical Transactions of the Royal Society of London* A, **344**, 59–67.

White, D.C., Davis, W., Nickels, J., King, J.D. & Bobbie, R.J. (1979) Determination of the sedimentary microbial biomass by extractable lipid phosphate. *Oceanologia* **40**, 51–62.

White, D.C., Smith, G.A., Gehron, M.J., Parker, J.H., Findlay, R.H., Martz, R.F. & Fredrickson, H.L. (1983) The ground water aquifer microbiota: biomass, community structure and nutritional status. *Developments in Industrial Microbiology* **24**, 201–11.

8

The behaviour of radionuclides in the coastal and estuarine environments of the Irish Sea

P. McDonald and S.R. Jones

8.1 Introduction

The focus of this chapter is on the behaviour of radionuclides in the coastal and intertidal sediments of the Irish Sea, in particular those in the north-east. A review of previous research concerned with the fate of radionuclides in and around the Irish Sea is presented along with some recent research the authors have performed.

8.1.1 Boundaries of the Irish Sea

The Irish Sea is a semi-enclosed body of water, bounded by the eastern coast of Ireland and the coasts of north Wales, north-west England and south-west Scotland. It connects with the Atlantic Ocean through St George's Channel in the south and the North Channel to the north. Water enters the Irish Sea both through St George's Channel and the North Channel, although the former dominates, and leaves via the North Channel (Howorth, 1984; Dickson, 1987). The flowrate of water out through the North Channel is about $5 \times 10^9\,m^3\,d^{-1}$ and the residence time of the water in the northern Irish Sea is about one year (Pentreath *et al.*, 1984).

In relation to radionuclide behaviour, the distribution of bed sediments according to grain size is of particular importance. Figure 8.1 shows the disposition of bed sediments according to the grain size description of Pantin (1977, 1978). The two main areas of muddy sediments, one extending southwards from St Bees Head, Cumbria, towards Morecambe Bay and the other lying between the Isle of Man and the north-east coast of Ireland, are of particular significance.

The dynamic behaviour of bed sediments has been the subject of much research and debate. Some authors (Mauchline, 1980; Pantin, 1991) consider that both the main areas of muddy sediment are actively accreting whilst others (Kirby *et al.*, 1983; Kershaw & Young, 1988) consider that the net accumulation rate is zero or very low (about 0.1 mm y^{-1}). Despite this uncertainty, the sediment is by no means static. Extensive vertical sediment mixing occurs through the action of biota,

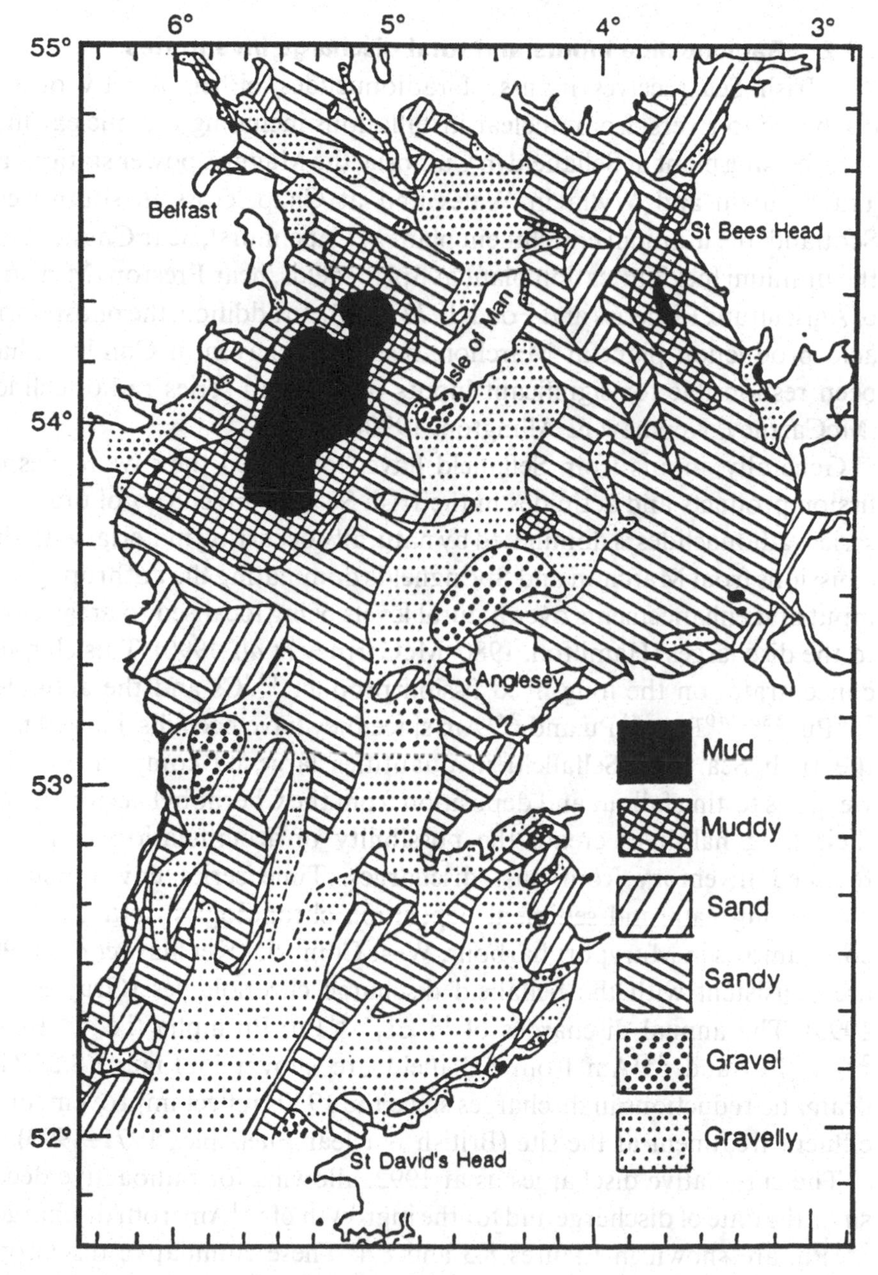

Figure 8.1 Distribution of bed sediments by grain size (Pantin, 1977, 1978).

principally the Echurian worm *Maxmülleria lankesteri* and the Thalassinid shrimp *Callinassa subterranea* (Kershaw *et al.*, 1984; Swift & Kershaw, 1986). Sediment resuspension arises from localised erosion, bioturbation and human activities such as trawling, and is balanced on average by deposition.

8.1.2 Radionuclide inputs and total discharge inventories

The Irish Sea receives inputs of radionuclides, either directly or via discharge to rivers, from nuclear installations including the nuclear fuel reprocessing plant at Sellafield in Cumbria, the Magnox power stations at Trawsfynydd and Wylfa in Wales and at Chapelcross in south-west Scotland, the uranium enrichment plant at Capenhurst, near Chester and the uranium fuel fabrication plant at Springfields, near Preston (Ministry of Agriculture, Fisheries and Food, 1972–1993). In addition, the phosphoric acid production plant at Marchon, near Whitehaven in Cumbria, has been responsible for significant inputs of uranium series radionuclides (McCartney, Kershaw & Allington, 1990).

Generally, inputs from Sellafield have dominated the inventories of fission products and actinides in the Irish Sea; the inventory of uranium series radionuclides is dominated by naturally occurring material with the emissions from Marchon and Springfields dominating the anthropogenic input and enhancements over natural levels being localised to areas close to the discharges (Hamilton, 1980; McCartney *et al.*, 1990). This chapter concentrates on the long lived fission product ^{137}Cs and the actinides ^{238}Pu, ^{239,240}Pu, ^{241}Pu and ^{241}Am since they have been discharged into the Irish Sea from Sellafield in quantities large by comparison with weapons testing fallout and deposition from the Chernobyl accident, and their long half-lives create the possibility of accumulation and/or a retained inventory from past discharges. Time series environmental monitoring data and sediment deposition chronologies obtained from core samples in Maryport harbour, West Cumbria (Kershaw *et al.*, 1990) are consistent with the Sellafield discharge chronologies (Gray *et al.*, 1995). The annual discharges of ^{137}Cs, ^{241}Pu, Pu alpha (i.e. ^{238}Pu + ^{239,240}Pu) and ^{241}Am from Sellafield are shown in Figure 8.2. The dramatic reductions in discharges since the 1970s reflect improvement in effluent treatment at the site (British Nuclear Fuels plc., 1971–1993).

The cumulative discharges as at 1992, allowing for radioactive decay since the date of discharge and for the ingrowth of ^{241}Am from discharged ^{241}Pu, are shown in Figures 8.3 and 8.4. These cumulative discharges represent the current global inventories resulting from the discharges; the actual inventories in the Irish Sea are lower, reflecting losses from the Irish Sea due to transport processes as later discussed.

8.2 Radionuclide behaviour: sediment–water interactions, sediment behaviour

The transport of radionuclides within the Irish Sea and further afield, and the development of the total inventory of the Irish Sea as a function of time, are determined by interactions of the radionuclides with the water

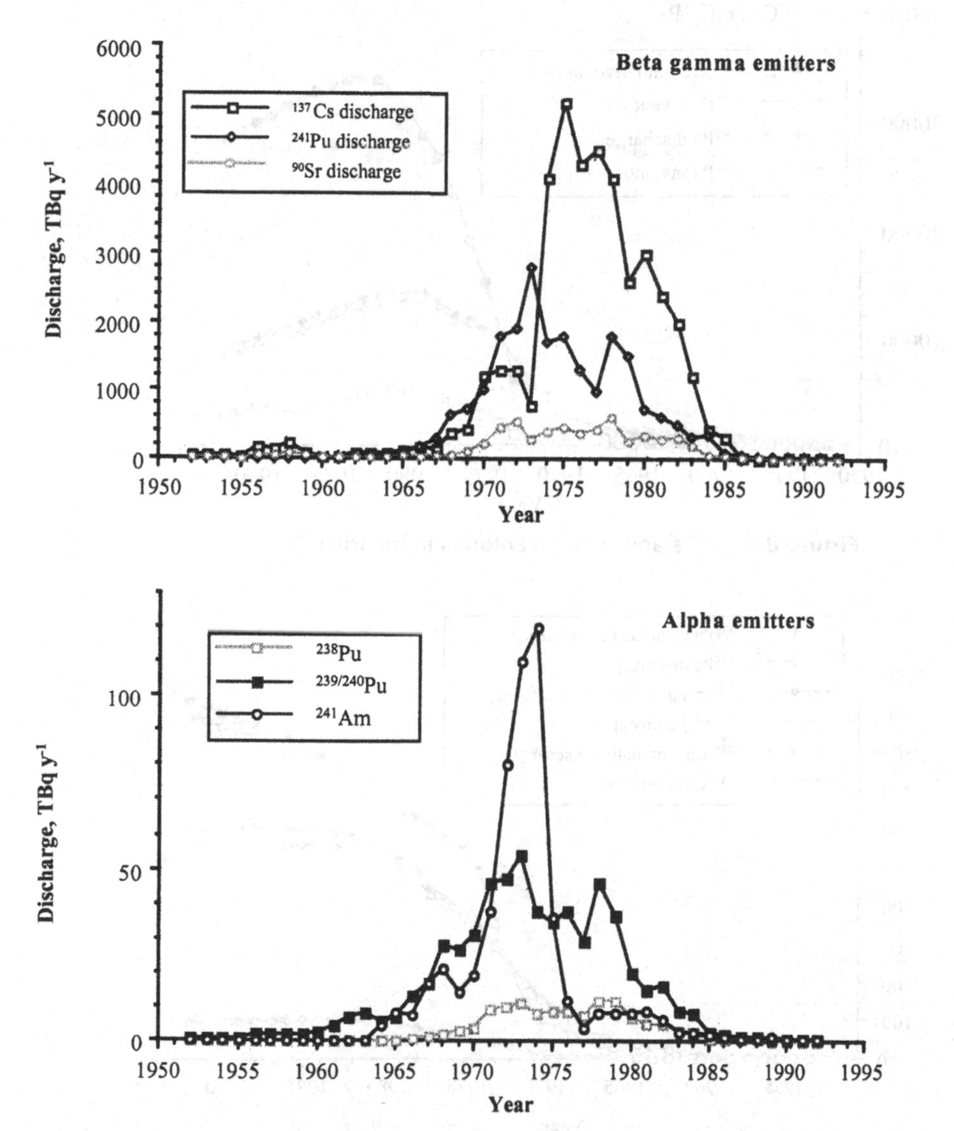

Figure 8.2 Discharges to the Irish Sea from Sellafield (British Nuclear Fuels plc., 1971–1993).

and sediment phases and by the transport processes affecting sea water and sediments. Caesium exhibits straightforward chemical behaviour as a metal from Group I, existing only in a single oxidation state. The environmental chemistry of the actinides is more complex as they can exist in a number of different oxidation states. Thus, plutonium can be present as Pu(III), Pu(IV), Pu(V) or Pu(VI). $^{239,240}Pu$ present in filtered sea water from the Irish Sea contains a mixture of these oxidation states

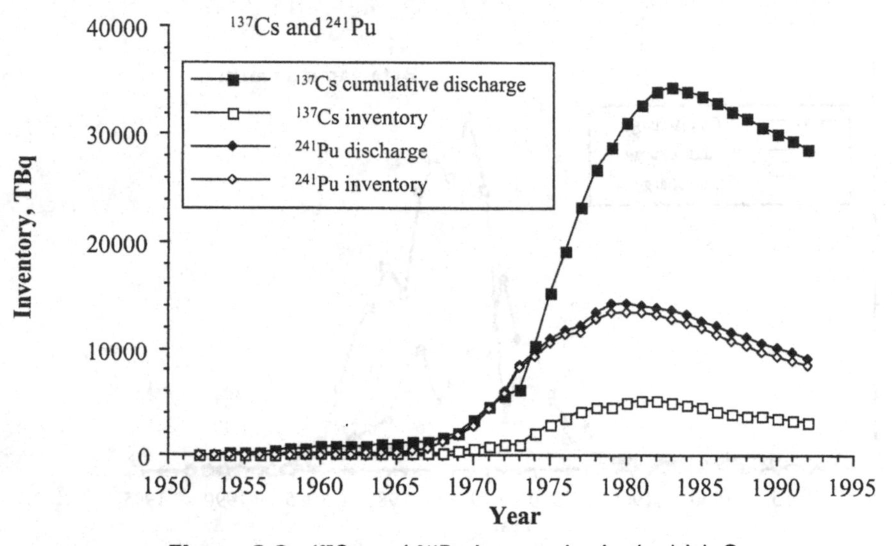

Figure 8.3 ^{137}Cs and ^{241}Pu inventories in the Irish Sea.

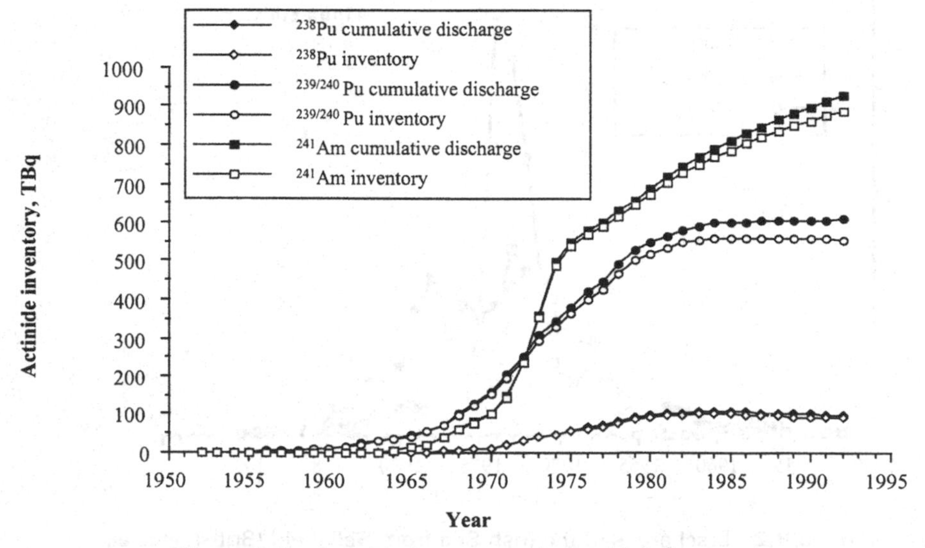

Figure 8.4 Actinide inventories in the Irish Sea.

(Nelson & Lovett, 1978; Pentreath, Harvey & Lovett, 1985) with more than 90% in the oxidised forms (V + VI) and Pu(V) probably predominating (Nelson & Orlandini, 1979). Less is known about the environmental chemistry of ^{241}Am but it appears to exist almost entirely as Am(III).

The most important aspect of radionuclide behaviour affecting transport and subsequent distribution is their propensity to adsorb onto sediments. This is conventionally characterised by reference to an equilibrium distribution coefficient, K_D, between sediment and water phases. Typical values for this coefficient are shown in Table 8.1 (Pentreath *et al.*, 1984; International Atomic Energy Agency (IAEA), 1985; Kershaw *et al.*, 1986; Howorth & Kirby, 1988). The coefficients given in Table 8.1 are the ratio of total radionuclide concentration in the sediment phase to that in the water phase, and may mask differing behaviour according to oxidation state. Thus, K_D values for Pu(III + IV) are of the order 10^6 whereas those for Pu(V + VI) are less than 10^4 (Nelson & Lovett, 1978). Accordingly, different oxidation states of Pu predominate in water and sediment phases and the apparent K_D values may vary with distance from the discharge point due to preferential scavenging of the reduced forms by sedimentation (Sholkovitz, 1983; Pentreath *et al.*, 1985).

Caesium is said to behave 'conservatively'; that is, the bulk of the radionuclide inventory is associated with the water phase and so transport processes are dominated by the bulk movement of sea water. Plutonium and americium, on the other hand, behave 'non-conservatively'; the bulk of their inventory is associated with sediments and the transport processes affecting sediments are very important to their behaviour. The proportion of each nuclide present in the water column as suspended particulate is a simple function of K_D value and suspended sediment load (Sholkovitz, 1983), as indicated in Table 8.2. Thus, for the full range of sediment loadings, water column inventories of ^{137}Cs are dominated by

Table 8.1 Sediment/water distribution coefficients (Pentreath *et al.*, 1984; IAEA, 1985; Kershaw *et al.*, 1986; Howorth & Kirby, 1988)

Radionuclide	Distribution coefficient (K_D)
^{137}Cs	10^3
^{239,240}Pu	3×10^5
Pu{III,IV}	10^6
Pu{V,VI}	10^4
^{241}Am	2×10^6

material in solution whereas those for ^{241}Am are dominated by material adsorbed to suspended sediments; water column inventories of ^{239,240}Pu are also dominated by suspended sediments at loadings above about 4 mg l^{-1}.

The adsorption of radionuclides to sediments is strongly affected by particle size and mineralogy (Aston, Assinder & Kelly, 1985; Ramsay & Raw, 1987; Livens & Baxter, 1988a). In particular, fine grained material accumulates higher concentrations of all radionuclides and the presence of clay minerals also enhances adsorption.

The net effect of the above processes is that the bulk of the Cs inventory is transported with water movements with a small fraction being adsorbed to suspended particulate and deposited in bed sediments, whereas a large fraction of the Pu and Am inventory is adsorbed to suspended sediments and deposited in the bed sediments of the eastern Irish Sea. The behaviour of Pu and Am deposited in bed sediments is then determined by the processes of sediment mixing (in which bioturbation is very important), resuspension, and remobilisation of adsorbed radionuclides into the solution phase.

8.2.1 Intertidal areas: beaches, harbours, estuaries, saltmarshes

Because of the strong inverse association between sediment grain size and radionuclide concentrations, it is the areas in which finer sediments accumulate that are of greatest significance. The disposition of coastal sediment types is shown in Figure 8.5. Exposed beaches exhibit coarser sediments, typically sand or shingle, because of the energy input from wave action. The radionuclide concentrations in these areas are generally low; however, because wave action mixes radionuclides to a substantial depth and because they represent a large proportion of the intertidal area,

Table 8.2 Fraction of total water column inventory carried by suspended sediment (Sholkovitz, 1983)

Nuclide	K_D	Fraction present as suspended sediment at stated sediment loading		
		1 mg l^{-1}	10 mg l^{-1}	100 mg l^{-1}
^{137}Cs	1×10^3	0.001	0.01	0.09
^{239,240}Pu	3×10^5	0.23	0.75	0.97
^{241}Am	2×10^6	0.67	0.95	0.99

they contain a large proportion of the total inventory of radionuclides in the Irish Sea (Eakins *et al.*, 1988, 1990; Garland *et al.*, 1989; Carpenter *et al.*, 1991).

Estuaries provide areas of lower energy where the finer grained sediments (silts and muds) can deposit. The estuary at Ravenglass, which is some 9 km from Sellafield and formed at the confluence of the rivers Irt, Mite and Esk, has been extensively studied by many researchers (e.g. Hetherington, 1976; Aston & Stanners, 1982a,b,c; Hamilton & Clarke, 1984; Aston *et al.*, 1985; Assinder, Kelly & Aston, 1985; Burton, 1986). Measurements have also been reported in estuaries from south-west Scotland to north Wales (Aston *et al.*, 1981; MacKenzie & Scott, 1982, 1993; McDonald *et al.*, 1992; Jones *et al.*, 1994).

Estuaries are complex and dynamic environments, with sediments subject to both erosion and deposition at different times and to varying water chemistry, particularly in salinity, pH and suspended sediment load. These all influence the behaviour of radionuclides (Clifton & Hamilton, 1982; Kelly *et al.*, 1991). A single estuary is likely to contain sedimentary deposits ranging from coarse sands to fine muds with a correspondingly large range of radionuclide concentrations. Most actively

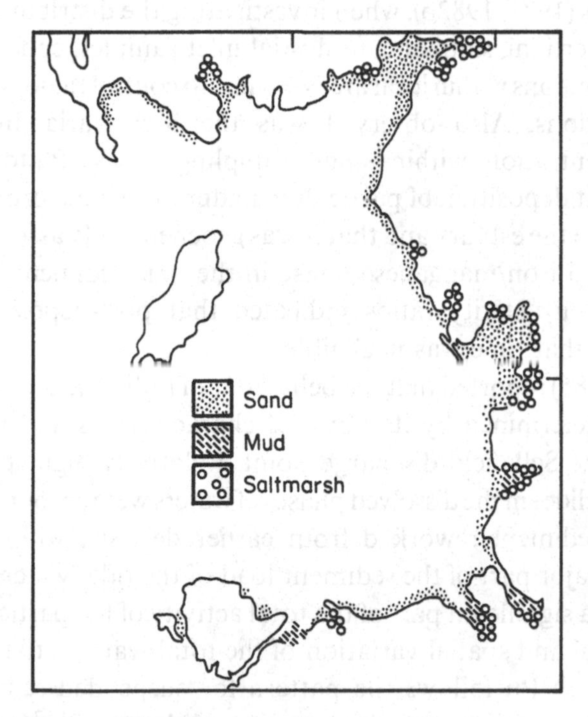

Figure 8.5 Coastal sediments types.

accreting sediment deposits also show a variation in radionuclide content with depth which can be correlated with the chronology of discharges from Sellafield and used to yield estimates of sedimentation rate (e.g. Aston & Stanners, 1979). However, the interpretation of such profiles is complicated by the fact that sedimentation rates at a particular site are unlikely to be constant with time, and may indeed vary from net erosion to net deposition (Clifton & Hamilton, 1982; MacKenzie & Scott, 1993).

Most of the estuarine conditions discussed above have been derived through study of radionuclides in the estuary at Ravenglass. However, because of the attention given to this estuarine area regarding radionuclide behaviour in an intertidal environment it is appropriate to provide more details of such research here.

Hamilton & Clarke (1984), in describing the recent sedimentation history at the Esk Estuary, found that the central regions of the estuary sediments were being deposited at a rate of 3 to 4 cm y^{-1}; along the banks lower rates of between 1 and 2 cm y^{-1} were found, while the saltmarsh proper was probably emerging at a rate of ~ 0.5 cm y^{-1}. These differences of sedimentation rate were directly related to grain size of sediments and biological activity, the higher rates being associated with the deposit of coarse sands and the lower rates being associated with deposition of fine silts.

Aston & Stanners (1981, 1982b), when investigating the distribution of radionuclides in intertidal, saltmarsh and tidal inlet samples, concluded that grain size distributions within local areas seemed to control radionuclide activity concentrations. Also observed was the high variability of radionuclide concentration within small sampling areas. Plutonium studies revealed that deposition of particulate material was the principal control of its input to the estuary and that it was geochemically associated with the non-detrital iron/manganese phase in the oxic sediment layer. Also $^{238}Pu/^{239,240}Pu$ activity ratios indicated that post-depositional migration of Pu at that time was negligible.

Assinder *et al.* (1985) reported that the behaviour of radionuclides in the Esk Estuary was determined by its physical characteristics and by its location close to the Sellafield discharge point. Relatively high specific activities of radionuclides in the dissolved phases of waters were encountered. They found that sediment reworked from earlier deposits within the estuary formed a major part of the sediment load of the tidal waters and therefore provided a significant part of the total activity of the particulate phase. The temporal and spatial variation of the total water activity for radionuclides such as Pu follows the pattern for suspended sediment concentration whereas, for conservative radionuclides (e.g. ^{137}Cs), the distribution of total water activity resembles that for salinity. Assinder *et al.* (1985) also reported that the particulate phase was dispersed more effectively within the estuary than the dissolved phase leading to proportionally greater contamination of the upper reaches at high water and over most of the estuary at low water.

Detailed investigations on the effect of low pH and salinity (Burton, 1986) showed that at low tide and when fresh water was backed up by the incoming tide, remobilisation of actinides into solution occurred in the estuary. These actinides, however, could then be readsorbed by the inactive river sediment load and redispersed around the estuary. This activity, however, may be lost from the estuary with the suspended sediment load on the ebb tide.

Although Sellafield discharges have been greatly reduced since 1980, Ravenglass is still a focus of attention due to the redispersion of historically deposited radionuclides and the intertidal processes that influence them.

Most estuaries, including Ravenglass, contain both exposed sediments and saltmarsh, which consists of sediment covered by salt tolerant plant species such as *Spartina anglica*, *Puccinellia maritima*, *Agrostis stolonifera*, *Festuca rubra* and others. Saltmarshes build up by continuing accretion of sediment, mostly as a consequence of the trapping of suspended sediment by vegetation during tidal inundations; the growth rate of the marsh decreases as the elevation above mean sea level increases, because the frequency of inundation decreases (Jones *et al.*, 1994). A saltmarsh will typically show a gradation from seaward to landward ends, the landward end being older and higher above mean sea level. Vegetation also shows a gradation with the less salt tolerant species limited to the higher areas where inundation is less frequent, and the diversity of vegetation greater. Saltmarsh vegetation is an effective trap for the deposition of fine sediment, so that saltmarshes usually contain silts and muds rather than coarser grained material, and so contain the highest concentrations of radionuclides in the estuary (Emptage & Kelly, 1990; Jones *et al.*, 1994). Saltmarsh sediments are stabilised by the vegetation and consequently often show well developed profiles of radionuclides with depth with subsurface maxima which correspond to the deposition of sediment at the time of maximum discharges from Sellafield (Bonnet, Appleby & Oldfield, 1988; Jones *et al.*, 1994). The depth of penetration of radionuclides into the profile is usually determined by the sedimentation rate over the period of highest discharges from Sellafield, i.e. 1965–1985. Areas of marsh which were in active growth at this time can have radionuclide profiles extending to 1 m or more in depth; areas of marsh which were mature at this time may have essentially all the radionuclides in the upper 0.2 m or less.

8.2.2 Radionuclide loss rates and total sediment inventories

Of the total available material (i.e. cumulative discharges corrected for radioactive decay and/or ingrowth) the fractions retained in the Irish Sea for ^{137}Cs, ^{239,240}Pu and ^{241}Am are approximately 10%, 90% and 95% respectively (MacKenzie *et al.*, 1994). The inventories for ^{137}Cs, ^{238}Pu, ^{239,240}Pu and ^{241}Pu are currently reducing at rates of 160, 1.5, 0.8 and 430 TBq y^{-1} respectively, as a result of advection through the North Channel and radioactive decay exceeding the current input rates from Sellafield, whilst the ^{241}Am inventory is increasing at 12 TBq y^{-1} largely because of ingrowth from the remaining inventory of ^{241}Pu.

Total inventories of ^{239,240}Pu and ^{241}Am in the Irish Sea have also been estimated from environmental measurements, particularly in sediments. Pentreath *et al.* (1984) have assessed the distribution in seabed sediments from surveys made in 1977 and 1978 as shown in Figure 8.6. They concluded that the total sediment inventory amounted to 280 TBq of

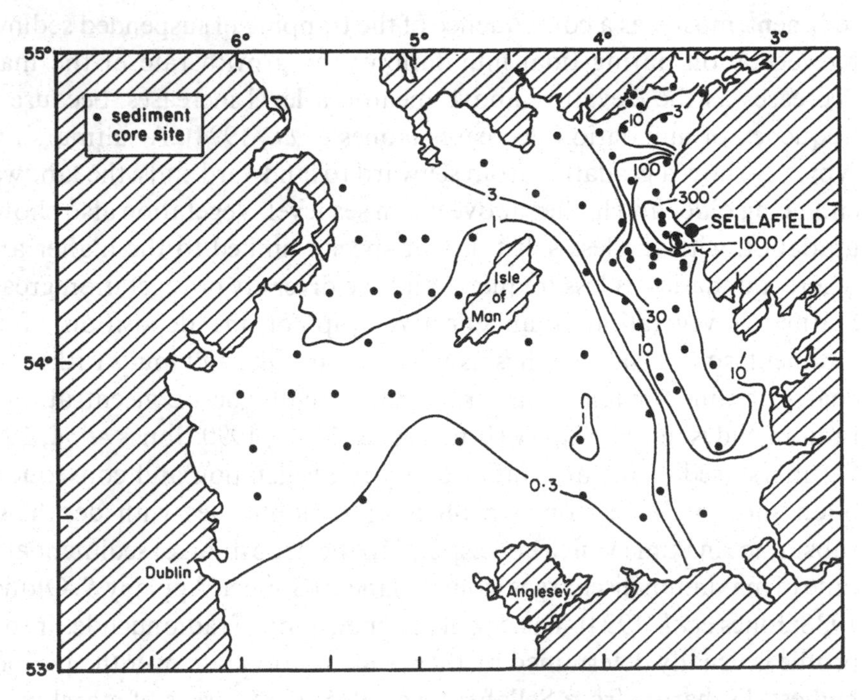

Figure 8.6 Plutonium inventory (kBq m^{-2}) in top 0.3 m of bed sediments (Pentreath *et al.*, 1984).

^{239,240}Pu and 340 TBq of ^{241}Am, most of which was contained in a coastal strip some 30 km wide extending from Kirkudbright Bay in the north to the Ribble Estuary in the south. This strip corresponds to an area of the seabed with a high proportion of fine grained sediment (Figure 8.1).

The intertidal areas on the north-west coast of England and the south-west coast of Scotland also contain significant inventories of radionuclides associated with sediments. Surveys made in the 1980s (Eakins *et al.*, 1988, 1990; Garland *et al.*, 1989; Carpenter *et al.*, 1991) have indicated total inventories of 36 TBq of ^{239,240}Pu and 50 TBq of ^{241}Am in the intertidal area between south-west Scotland and the coast of north Wales, as summarised in Table 8.3.

The inventory summarised above may be an underestimate, because the surveys covered largely exposed mud and sand and, although the total area covered by saltmarsh is lower (Carpenter *et al.*, 1991), the inventories per unit area for saltmarsh can be higher (Oldfield *et al.*, 1993). The measured inventories may be compared with the inventories estimated for other compartments of the Irish Sea as summarised in Table 8.4.

It can be seen that about 70% of the calculated inventory of ^{239,240}Pu and ^{241}Am are accounted for by the measurements. It seems unlikely that underestimate of the loss through the North Channel could account for the discrepancy. It is likely that Pu and Am penetrating below the 0.3 m core depth sampled by Pentreath *et al.* (1984), and perhaps additional

Table 8.3 Inventories of actinides in the intertidal areas of north-west England and south-west Scotland (Eakins *et al.*, 1988, 1990; Garland *et al.*, 1989; Carpenter *et al.*, 1991)

	Radionuclide inventory (TBq)		
Area	^{238}Pu	^{239,240}Pu	^{241}Am
Wigtown Bay	0.4	1.9	2.9
Inner Solway*	1.0	4.0	6.0
Cumbrian coast	1.2	5.7	8.9
Morecambe Bay	2.2	10.8	15.6
Lancashire coast	1.8	11.6	12.7
Ribble Estuary	0.2	1.1	3.6
Mersey Estuary	0.02	0.02	0.2
Dee Estuary	0.2	0.9	1.0
Total	7	36	51

*Estimated

material in the intertidal areas not yet comprehensively surveyed, will account for the 'missing' Pu and Am.

8.2.3 Geochemical associations of radionuclides

It is now well established that particle transport is the dominant mechanism of supply of Sellafield-derived radionuclides to the Solway region (MacKenzie, Scott & Williams, 1987; McDonald *et al.*, 1990). Onshore transfer of this sediment from further south in the Irish Sea is obvious upon consideration of the high sedimentation rates in these areas and the fact that there is no significant fluvial input (McDonald, Cook & MacKenzie, 1993). It is of importance to establish whether changes in speciation and availability of radionuclides occur upon transport from the highly saline, alkaline and well oxygenated marine environment to terrestrial areas where the ground water is brackish, pH values range from slightly alkaline to slightly acidic and oxygen depleted conditions can occur at depth due to degradation of organic matter. Described below are some aspects of recent studies of the mechanism of radionuclide transport to the Solway Coast and of the geochemical behaviour of radionuclides in Solway floodplain sediments.

A transect of five surface sediments (cores A–E, Table 8.5) from the north-east Irish Sea revealed the specific activities of ^{137}Cs, ^{241}Am and Pu isotopes to be of the same magnitude over a 60 km transect (except for core C, in which activities were an order of magnitude lower) (McDonald *et al.*, 1990). The size fractionation results (Table 8.6) revealed an inverse correlation between particle size and radionuclide concentration, confirming the well documented preferential association of radionuclides

Table 8.4 Actinide inventory estimates for the Irish Sea (referred to 1977/78) (Willans, Smith & Jones, 1994)

Compartment	Inventory, TBq	
	^{239,240}Pu	^{241}Am
Seabed sediments	280	340
Intertidal sediments	36	51
Water column	5	5
Total measured	320	390
Discharge less loss through North Channel	450	600

with finer size fractions (Hetherington, 1976). These data indicate that the pool of radionuclides in the surface offshore sediment of the north-east Irish Sea is spatially well mixed over the area studied (cf. cores A and E) and that radionuclide concentrations are subject to greater control by particle size composition rather than distance from Sellafield (McDonald *et al.*, 1990).

The geochemical behaviour of Cs and Pu in Solway floodplain cores, collected from Southwick Water, south-west Scotland, has been studied by Allan *et al.* (1991) by employing a sequential leaching scheme (Cook *et al.*, 1984; Livens & Baxter, 1988b; McDonald *et al.*, 1990). This scheme incorporates solutions of calcium chloride (readily available sites), acetic acid (specific adsorption sites), tetra-sodium pyrophosphate (organically associated), ammonium oxalate/oxalic acid, (Fe and Mn secondary minerals), dilute nitric acid (dilute acid soluble sites) and nitric acid/hydrofluoric acid (residual fraction). This technique has been subject to

Table 8.5 Surface radionuclide concentrations of north-east Irish Sea sediments (Bq kg^{-1} ± 2σ counting error) (McDonald *et al.*, 1990)

Core	Depth (m)	Distance from Sellafield (km)	^{137}Cs	^{238}Pu	^{239,240}Pu	^{241}Am
A	37	14	596 ± 26	62 ± 7	328 ± 32	392 ± 26
B	35	20	229 ± 10	46 ± 7	205 ± 26	205 ± 14
C	40	29	69 ± 4	7 ± 1	38 ± 3	55 ± 4
D	45	49	262 ± 12	49 ± 6	233 ± 24	236 ± 15
E	15	63	462 ± 22	61 ± 20	297 ± 26	309 ± 18

Table 8.6 Particle size fractionation of north-east Irish Sea surface sediments (McDonald *et al.*, 1990)

Core	% Organic carbon	% Coarse sand (> 160 μm)	% Fine sand (50–160 μm)	% Coarse silt (20–50 μm)	% Fine silt (2–20 μm)	% Clay (< 2 μm)
A	8.0	2.9	46.0	14.8	22.9	13.4
B	3.6	27.3	53.3	4.8	3.2	9.3
C	2.0	84.3	10.3	0	0	5.4
D	2.9	68.1	16.1	2.3	4.3	11.4
E	5.2	3.2	68.3	11.7	3.3	13.4

Particle size distributions calculated from total mineral content. Organic fraction determined from sample weight loss at 500°C.

recent general criticism (Khoeboian & Bauer, 1987) but remains a valuable method of investigating the ease of removal of radionuclides and other metals from soils and sediments, and in the present approach sequential leaching is regarded as selectively dissolving radionuclides in operationally defined, notionally discrete components of the sample.

It is evident from Table 8.7 that, independent of the depth (or age) of the sediment, negligible quantities of ^{137}Cs were dissolved by reagents selected to remove the available, exchangeable, organic bound and secondary iron/manganese bound components. In each case, at least 80% of the ^{137}Cs was tightly bound in the residual phase and between 11 and 15% was associated with the fraction extracted by 1MHNO_3. This is consistent with the observation by Allan *et al.* (1991) of a K_D of about 10^5 applicable to desorption of ^{137}Cs from this sediment.

Associations of ^{239,240}Pu in surface samples revealed 41% in the 'organic' phase, 39% in the iron/manganese extract and 16% in the residual fraction, with minor quantities in the other compartments (Table 8.8). A systematic variation in these associations was observed as a function of depth with the 'organic' component containing progressively less ^{239,240}Pu and the iron/manganese and residual fractions containing correspondingly more. In no instance was there any significant release of ^{239,240}Pu in the available or exchangeable extractions.

These trends, in conjunction with an observed decrease in organic content from 5% at the surface to 3% at depth, suggest that, as organic matter is degraded, some of the plutonium which it originally contained is released but is immediately taken up by other components of the sediment (i.e. Fe/Mn secondary minerals). This observation is compatible with

Table 8.7 Sequential ^{137}Cs leaching from Southwick Water floodplain (% distribution) (Allan *et al.*, 1991)

Depth (cm)	%Organic carbon	RA	EX	OR	Fe/Mn	DAS	RE
0–5	5.2	BDL	BDL	BDL	2	12	86
20–25	5.1	BDL	BDL	0.4	4	15	81
40–45	4.8	2	BDL	BDL	3	12	83
55–60	4.2	BDL	BDL	BDL	2	14	84
65–70	3.8	BDL	4	BDL	1	11	84

RA – readily available; EX – exchangeable; OR – organically bound; Fe/Mn – Fe/Mn secondary minerals; DAS – dilute acid soluble; RE – residual; BDL – below detection limits recent work on this sediment (Graham, Livens & Scott, 1993) showing that the abundance of functional groups in humic acids (in particular – COOH) decrease with depth and hence remove potential radionuclide binding sites.

The results of the sequential leaching experiments for plutonium in a Solway soil, an intertidal sediment and two Irish Sea sediments are presented in Table 8.9 (McDonald *et al.*, 1990). It is evident that, for all samples analysed, almost of all of the plutonium (> 98%) is associated in the more strongly bound fractions (organic, Fe/Mn secondary minerals and residual phases) consistent with the results for floodplain sediments. In the marine and intertidal sediments, the general order of geochemical association of plutonium is: Fe/Mn secondary minerals > organic > residual > specific adsorption > readily available. In the soil, the order of association is: organic > Fe/Mn secondary minerals > residual > readily available, specific adsorption. The soil sample was collected from just

Table 8.8 Sequential $^{239,240}Pu$ leaching from Southwick Water floodplain (% distribution) (Allan *et al.*, 1991)

Depth (cm)	%Organic carbon	RA	EX	OR	Fe/Mn	DAS	RE
0–5	5.2	BDL	BDL	41	39	2	16
20–25	5.1	BDL	0.3	43	40	0.5	16
40–45	4.8	BDL	1	24	51	2	22
55–60	4.2	BDL	BDL	34	46	1	19
65–70	3.8	BDL	BDL	17	57	2	24

RA – readily available; EX – exchangeable; OR – organically bound; Fe/Mn – Fe/Mn secondary minerals; DAS – dilute acid soluble; RE – residual; BDL – below detection limits

Table 8.9 Sequential $^{239,240}Pu$ leaching from a selection of Solway sediments and soil (% distribution, ± 1σ counting errors) (McDonald *et al.*, 1990)

Sample	% Organic carbon	RA	EX	OR	Fe/Mn	RE
Core A	8.0	< 0.5	0.50 ± 0.05	31 ± 3	47 ± 4	22 ± 2
Core E	5.2	< 0.05	0.70 ± 0.04	40 ± 1	37 ± 2	23 ± 2
Kippford	5.3	< 0.05	< 0.05	32 ± 2	56 ± 3	11 ± 1
Meickle-Richorn soil	8.3	0.10 ± 0.02	< 0.05	56 ± 3	32 ± 1	11 ± 1

RA – readily available; EX – exchangeable; OR – organically bound; Fe/Mn – Fe/Mn secondary minerals; RE – residual above the mean high water mark at a location subject to periodic tidal inundation, and the isotope activity ratio results confirm that the plutonium is from Sellafield waste. The plutonium associations, however, suggest a redistribution from the Fe/Mn secondary mineral and perhaps residual phases to the organic phase upon transition from the aquatic to the terrestrial environment.

Finally another area of interest is the 'mud-patch' off Sellafield acting as a source of radionuclides to the Irish Sea environment. From a core taken there in 1992, McDonald *et al.* (1993) found that the $^{137}Cs/^{241}Am$ activity ratio decreased systematically from 6.60 at a depth of 34–36 cm to 1.36 at the surface (Figure 8.7). While these variations are small relative to those of the annual discharge (7.0 to 1218), they indicate that, in addition to mixing (evident from the constancy of Pu/Am activity ratios down the core), there has been a loss of ^{137}Cs relative to ^{241}Am from this sediment. The results suggest that radiocaesium redissolution is significant to a depth of 25 cm or more which is in contrast to the results of McCartney *et al.* (1994) who estimated that 10 cm was the effective limit for this process. The major source of ^{137}Cs to the Irish Sea was predominantly from Sellafield. Although ^{137}Cs discharges are still performed, their magnitude compared to previous years is very small. The re-dissolution ^{137}Cs from historical deposits, however, is now a significant source of ^{137}Cs to the intertidal areas of eastern Irish Sea. Such a source term has to be considered when interpreting ^{137}Cs behaviour in these intertidal areas.

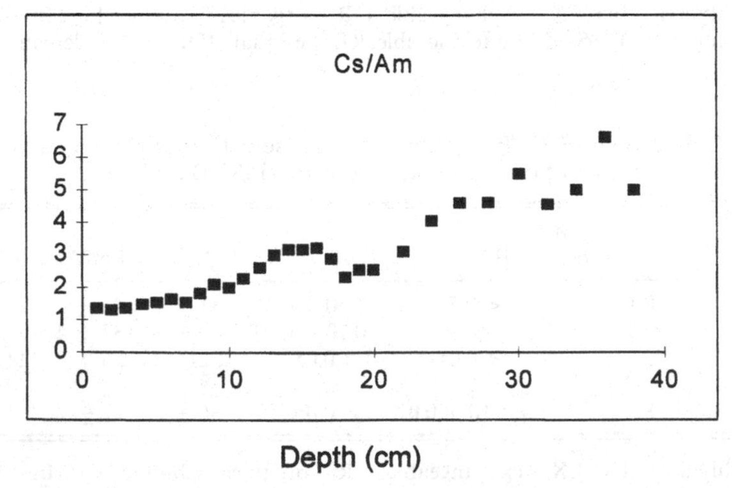

Figure 8.7 Activity ratio profile in Irish Sea sediment (McDonald, Cook & MacKenzie, 1993).

References

Allan, R.L., Cook, G.T., MacKenzie, A.B. & Pulford, I. (1991) Vertical distribution and geochemical associations of radionuclides in Solway Firth saltmarsh sediment. In *Heavy Metals in the Environment*, ed. J.G. Farmer, CEP Consultants Ltd Publishers, 445–8.

Assinder, D.J., Kelly, M. & Aston, S.R. (1985) Tidal variations in dissolved and particulate phase radionuclide activities in the Esk Estuary, England, and their distribution coefficients and particulate activity fractions. *Journal of Environmental Radioactivity* **2**, 1–22.

Aston, S.R. & Stanners, D.A. (1979) The determination of estuarine sedimentation rates by Cs-134/Cs-137 and other artificial radionuclide profiles. *Estuarine and Coastal Marine Science* **9**, 529–41.

Aston, S.R. & Stanners, D.A. (1981) Americium in intertidal sediments from the coastal environs of Windscale. *Marine Pollution Bulletin* **12**, 149–53.

Aston, S.R. & Stanners, D.A. (1982a) Gamma emitting fission products in surface sediments of the Ravenglass Estuary. *Marine Pollution Bulletin* **13**, 135–8.

Aston, S.R. & Stanners, D.A. (1982b) Local variability in the distribution of Windscale fission products in estuarine sediments. *Estuarine and Coastal Shelf Science* **14**, 167–74.

Aston, S.R. & Stanners, D.A. (1982c) The transport to, and deposition of, americium in intertidal sediments of the Ravenglass estuary and its relationship to plutonium. *Environmental Pollution* **3**, 1–10.

Aston, S.R., Assinder, D.J. & Kelly, M. (1985) Plutonium in intertidal coastal and estuarine sediments in the northern Irish Sea. *Estuarine and Coastal Shelf Science* **20**, 761–71.

Aston, S.R., Assinder, D.J., Stanners, D.A. & Rae, J.E. (1981) Plutonium occurrence and phase distribution in sediments of the Wyre estuary, north-west England. *Marine Pollution Bulletin* **12**, 308–14.

Bonnett, P.J.P., Appleby, P.G. & Oldfield, F. (1988) Radionuclides in coastal and estuarine sediments from Wirral and Lancashire. *Science of the Total Environment* **70**, 215–36.

British Nuclear Fuels plc. (1971–1993) Radioactive discharges and monitoring of the environment. Published annually, British Nuclear Fuels plc., Warrington, Cheshire.

Burton, P.J. (1986) Laboratory studies on the remobilisation of actinides from Ravenglass estuary sediment. *Science of the Total Environment* **52**, 123–45.

Carpenter, R.C., Burton, P.J., Strange, L.P. & Pratley, F.W. (1991) Radionuclides in intertidal sands from Morecambe Bay to the Dee Estuary. *Report AERE-R* 13803, AEA Technology, Harwell, Oxfordshire.

Clifton, R.J. & Hamilton, E.I. (1982) The application of radioisotopes in the study of estuarine sedimentary processes. *Estuarine and Coastal Shelf Science* **14**, 438–46.

Cook, G.T., Baxter, M.S., Duncan, H.J., Toole, J. & Malcolmson, R. (1984) Geochemical association of plutonium in the Caithness environment. *Nuclear Instruments and Methods in Physics* **223**, 517–22.

Dickson, R.R. (ed.) (1987) Irish Sea status report of the Marine Pollution Monitoring Management Group. *Aquatic Environment Monitoring Report, MAFF Directorate of Fisheries Research Lowestoft* **17**, 1–83.

Eakins, J.D., Morgan, A., Baston, G.M.N., Pratley, F.A., Yarnold, L.P. & Burton, P.J. (1988) Studies of environmental radioactivity in Cumbria. Part 8. Plutonium and americium in the intertidal sands of north-west England. *Report AERE-R*12061, AEA, Harwell.

Eakins, J.D., Morgan, A., Baston, G.M.N., Pratley, F.W., Strange, L.P. & Burton, P.J. (1990) Measurement of α-emitting plutonium and americium in the intertidal sands of West Cumbria, UK. *Journal of Environmental Radioactivity* **11**, 37–54.

Emptage, M. & Kelly, M. (1990) Gamma-dose rates in the Esk Estuary, *Department of the Environment Report DOE/RW/90/074.*

Garland, J.A., McKay, W.A., Cambray, R.S. & Burton, P.J. (1989) Man-made radionuclides in the environment of Dumfries and Galloway. *Nuclear Energy* **28**, 369–92.

Graham, M.C., Livens, F.R. & Scott, R.D. (1993) Associations of actinides with soil humic substances in saltmarsh sediments in south-west Scotland. In *Heavy Metals in the Environment*, eds. R.J. Allan and J.O. Nriagu, CEP Consultants Ltd Publishers, 239–42.

Gray, J., Jones, S.R. & Smith, A.D. (1995) Discharges to the environment from the Sellafield site, 1952–1992. *Journal of Radiological Protection* **15**, 99–131.

Hamilton, E.I. (1980) Concentration and distribution of uranium in *Mytilus edulis* and associated materials. *Marine Ecology Progress Series* **2**, 61–73.

Hamilton, E.I. & Clarke, K.R. (1984) The recent sedimentation history of the Esk Estuary, Cumbria, UK: the application of radiochronology. *Science of the Total Environment* **35**, 325–86.

Hetherington, J.A. (1976) The behaviour of plutonium nuclides in the Irish Sea. In *Environmental Toxicity of Aquatic Radionuclides*, eds. M.W. Miller and J.N. Stannard, 81–106. Ann Arbor Science Publishers Inc., Ann Arbor, Michigan.

Howorth, J.M. (1984) Currents in the eastern Irish Sea. *Oceanographic Marine Biology, Annual Review* **22**, 11–53.

Howorth, J.M. & Kirby, C.R. (1988) Studies of environmental radioactivity in Cumbria. Part 11 – Modelling the dispersion of radionuclides in the Irish Sea. *Report AERE-R11734*, AEA, Harwell.

IAEA (1985) Sediment K_Ds and concentration factors for radionuclides in the marine environment. *Technical report series no.* 247, International Atomic Energy Agency, Vienna.

Jones, S.R., Rudge, S.A., Murdock, R.N. & Johnson, M.S. (1994) The dynamics of vegetation contamination by radionuclides on a tide washed pasture in the Mersey estuary. *Water, Air and Soil Pollution* **75**, 265–75.

Kelly, M., Emptage, M., Mudge, S., Bradshaw, K. & Hamilton-Taylor, J. (1991) The relationship between sediment and plutonium budgets in a small macrotidal estuary: Esk Estuary, Cumbridge, UK. *Journal of Environmental Radioactivity* **13**, 55–74.

Kershaw, P.J. & Young, A. (1988) Scavenging of ^{234}Th in the eastern Irish Sea. *Journal of Environmental Radioactivity* **6**, 1–23.

Kershaw, P.J., Swift, D.S., Pentreath, R.J. & Lovett, M.B. (1984) The incorporation of plutonium, americium and curium into the Irish Sea seabed by biological activity. *Science of the Total Environment* **40**, 61–81.

Kershaw, P.J., Pentreath, R.J., Harvey, B.R., Lovett, M.B. & Boggis, S.J. (1986) Apparent distribution coefficients of transuranium elements in UK coastal waters. In *Application of Distribution Coefficients to Radiological Assessment Models*, eds. T.H. Sibley and C. Myttenaere, 227–87. Elsevier Applied Science Publishers, London and New York.

Kershaw, P.J., Woodhead, D.S., Malcolm, S.J., Allington, D.J. & Lovett, M.B. (1990) A sediment history of Sellafield discharges. *Journal of Environmental Radioactivity* **12**, 201–41.

Khoeboian, C. & Bauer, C.F. (1987) Accuracy of selective extraction procedures for metal speciation in model aquatic sediments. *Analytical Chemistry* **59**, 1417–23.

Kirby, R., Parker, W.R., Pentreath, R.J. & Lovett, M.B. (1983) Sedimentation studies relevant to low-level radioactive effluent dispersal in the Irish Sea. Part III. An evaluation of possible mechanisms for the incorporation of radionuclides into marine sediments. *Report from the Institute of Oceanographic Sciences*, Wormley, (178). 1–63.

Livens, F.R. & Baxter, M.S. (1988a) Particle size and radionuclide levels in some West Cumbrian soils. *Science of the Total Environment* **70**, 1–17.

Livens, F.R. & Baxter, M.S. (1988b) Chemical associations of artificial radionuclides in Cumbrian soils. *Journal of Environmental Research* **7**, 75–86.

MacKenzie, A.B. & Scott, R.D. (1982) Radiocaesium and plutonium in intertidal sediments from southern Scotland. *Nature* **299**, 613–16.

MacKenzie, A.B. & Scott, R.D. (1993) Sellafield radionuclides in Irish Sea sediments. *Environmental Geochemistry and Health* **15**, 173–84.

MacKenzie, A.B., Scott, R.D. & Williams, T.M. (1987) Mechanisms for northward dispersal of Sellafield waste. *Nature* **329**, 42–5.

MacKenzie, A.B., Scott, R.D., Allan, R.L., Ben Shaban, Y.A., Cook, G.T. & Pulford, I.D. (1994) Sediment radionuclide profiles: implications for the mechanisms of Sellafield waste dispersal in the Irish Sea, *Journal of Environmental Radioactivity* **23**, 39–69.

Mauchline, J. (1980) Artificial radioisotopes in the marginal seas of north western Europe. In *The North-west European Shelf Sea: the Sea Bed and the Sea in Motion. II. Physical and Chemical Oceanography and Physical Resources*, eds. F.T. Banner, M.B. Collins and K.S. Massie, 517–42. Elsevier. Amsterdam.

McCartney, M., Kershaw, P.J. & Allington, D.J. (1990) The behaviour of ^{210}Pb and ^{226}Ra in the eastern Irish Sea. *Journal of Environmental Radioactivity* **12**, 243–65.

McCartney, M. Kershaw, P.J., Woodhead, D.S. & Denoon, D.C. (1994) Artificial radionuclides in the surface sediments of the Irish Sea, 1968–1988. *Science of the Total Environment* **141**, 103–38.

McDonald, P., Cook, G.T., Baxter, M.S. & Thompson, J.C. (1990) Radionuclide transfer from Sellafield to south-west Scotland. *Journal of Environmental Radioactivity* **12**, 258–98.

McDonald, P., Cook, G.T., Baxter, M.S. & Thompson, J.C. (1992) The terrestrial distribution of artificial radioactivity in south-west Scotland. *Science of the Total Environment* **111**, 59–82.

McDonald, P., Cook, G.T. & MacKenzie, A.B. (1993) The dispersal of radionuclides in the Irish Sea. In *Heavy Metals in the Environment*, eds. R.J. Allan and J.O. Nriagu, CEP Consultants Ltd Publishers, 239–42.

Ministry of Agriculture, Fisheries & Food (1972–1993) Radioactivity in the coastal and surface waters of the UK. Published annually, MAFF Directorate of Fisheries Research, Lowestoft.

Nelson, D.M. & Lovett, M.B. (1978) Oxidation states of plutonium in the Irish Sea. *Nature* **276**, 599–601.

Nelson, D.M. & Orlandini, K.A. (1979) Identification of Pu(V) in natural waters. *Report ANL-79-65, part III*, 57–9, Argonne National Laboratory.

Oldfield, F., Richardson, N., Appleby, P.G. & Yu, L. (1993) ^{241}Am and ^{137}Cs activity in fine grained saltmarsh sediments from parts of the N. E. Irish Sea shoreline. *Journal of Environmental Radioactivity* **19**, 1–24.

Pantin, H.M. (1977) Quaternary sediments of the northern Irish Sea. In *Quaternary History of the Irish Sea*, eds. C. Kidson and M.J. Tooley, 27–54. Seel House Press, Liverpool.

Pantin, H.M. (1978) Quarternary sediments from the northeast Irish Sea: Isle of Man to

Cumbria. *Bulletin of the Geological Survey of Great Britain* **64**, 1–43.

Pantin, H.M. (1991) The sea-bed sediments around the United Kingdom. *Research Report of the British Geological Survey, Offshore Geological Services*, Keyworth, SB/90/1.

Pentreath, R.J., Harvey, B.R. & Lovett, M.B. (1985) Chemical speciation of transuranium nuclides discharged into the marine environment. In *Speciation of Fission and Activation Products in the Environment*, eds. R.A. Bulman and J.R. Cooper, 312–25, Elsevier, London.

Pentreath, R.J., Lovett, M.B., Jefferies, D.F., Woodhead, D.S., Talbot, J.W. & Mitchell, N.T. (1984) The impact on public radiation exposure of transuranium nuclides discharges in liquid wastes from fuel element reprocessing at Sellafield, UK. In *Radioactive Waste Management*, Proc. Symp., Seattle, 1983, 5, 315–329, International Atomic Energy Agency, Vienna.

Ramsay, J.D.F. & Raw, G. (1987) Physical characteristics of colloids and particulates in coastal sediments and seawater. *Report AERE R-12086*, AEA, Harwell.

Sholkovitz, E.R. (1983) The geochemistry of plutonium in fresh and marine water environments. *Earth Science Reviews* **19**, 95–161.

Stather, J.W., Dionian, J., Brown, J., Fell, J.P. & Muirhead, C.R. (1986) The risks of leukaemia and other cancers in Seascale from radiation exposure. *NRPB-R171 Addendum*, HMSO, London.

Swift, D.J. & Kershaw, P.J. (1986) Bioturbation of contaminated sediments in the north-east Irish Sea. International Council for the Exploration of the Sea, Copenhagen, *CM 1986/E: 18*.

Willans, S.M., Smith, A.D. & Jones, S.R. (1994) Development validation of the Sellafield Environmental Assessment Model (SEAM). *IRPA Regional Congress on Radiological Protection, Portsmouth 6–10 June 1994. Proceedings of the 17th Congress*, 243–6. Nuclear Technology Publishing.

9

The sorption of hydrophobic pyrethroid insecticides to estuarine particles: a compilation of recent research

J.L. Zhou and S.J. Rowland

9.1 Introduction

Intertidal sediments, especially fine-grained estuarine sediments, may be a net sink, or at least a transient repository of hydrophobic organic chemicals (HOC) from a number of sources. HOC have a strong, though variable, affinity for sediment minerals, especially for clays coated with natural organic matter (OM). Measurement of this adsorptive capacity is thus an important goal for geochemists wishing to study the budget and fate of HOC in the intertidal environment.

Although a considerable literature of adsorption measurements exists, the cocktail of HOCs which reaches estuarine waters is so diverse and originates from so many sources (including from municipal discharges, from effluents from industrial works sited on estuaries, and from riverine inputs) that inevitably there are important gaps in our knowledge. Included in this is information about a number of important agrochemical HOCs which may enter estuaries, via riverine transport. This chapter focuses particularly on one group of HOCs (pyrethroids), the chapter by Turner and Tyler discusses the behaviour of some other groups.

The pyrethroid agrochemicals represent a quarter of the world insecticides market and are widely used in the UK, continental Europe, the USA and elsewhere as household, public health and agricultural insecticides (Hill, 1989; Perrior, 1993). The chemicals are described by the generic name 'pyrethroids' because they are all man-made analogues of the esters of the natural cyclopropanoic acid, chrysanthemic acid, found in the plant family pyrethrums, but in fact they exhibit a range of physico-chemical properties in the environment. For example, the high octanol–water partition constants (K_{ow} 1 260 000 to 10 000 000; Figure 9.1) of the pyrethroids reflect a high affinity for organic matter such as that found in soils. Indeed, this adsorption onto (or partitioning into) organic matter may to a large extent immobilise the chemicals at the site of application. This restriction on transport is important since pyrethroids have a high toxicity to some aquatic organisms (Hill, 1985). Nonetheless small but measurable concentrations of pyrethroids are found in river waters (NRA, 1992), possibly due to physical transport of the soil-sorbed chemicals, for example, during periods of heavy rainfall. Once in river water, sorption

Permethrin, $K_{ow} = 1\ 260\ 000$

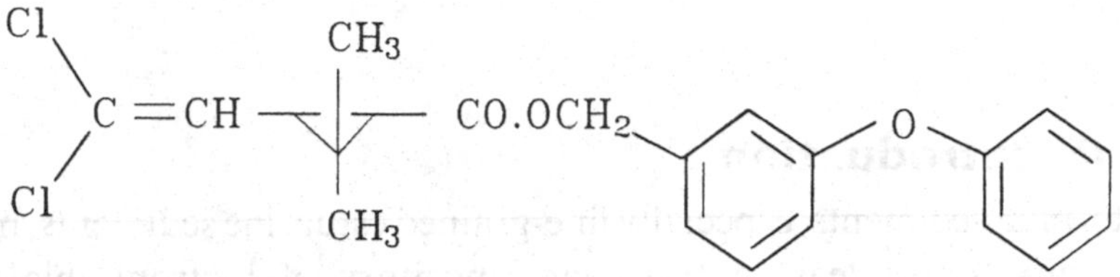

Tefluthrin, $K_{ow} = 3\ 160\ 000$

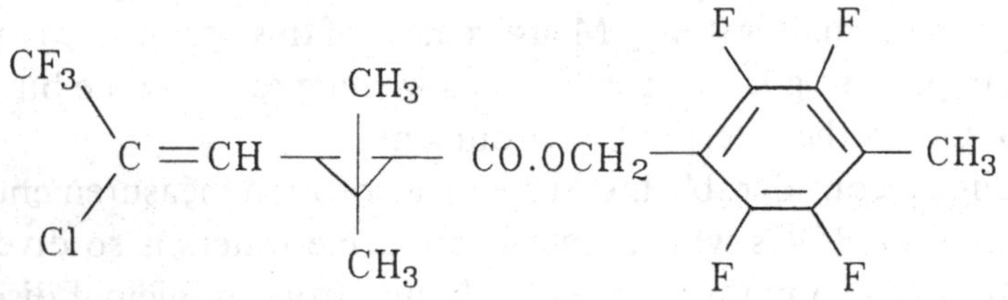

Cypermethrin, $K_{ow} = 4\ 000\ 000$

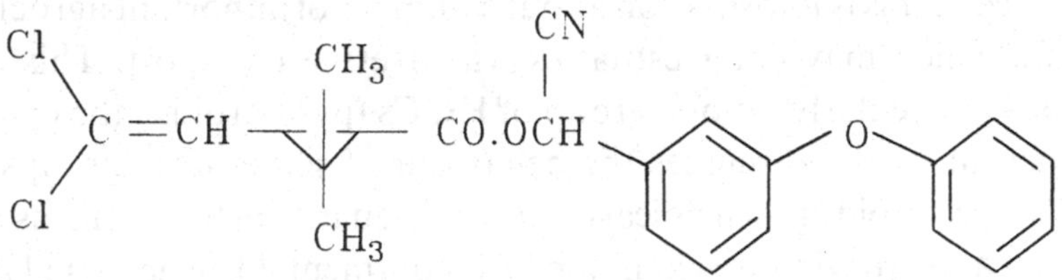

Lambdacyhalothrin, $K_{ow} = 10\ 000\ 000$

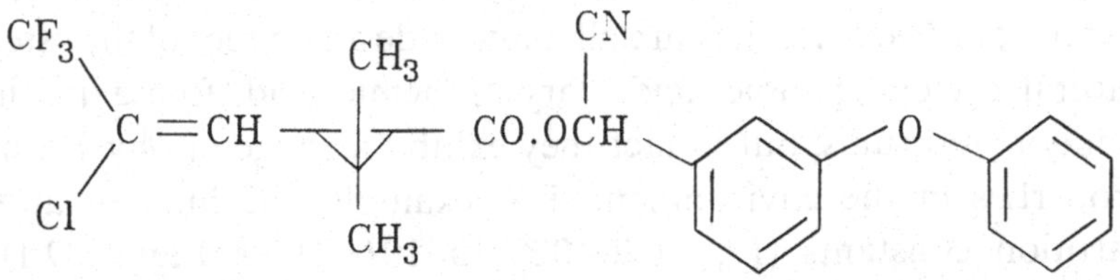

Figure 9.1 Chemical structure of the four pyrethroids and their respective K_{ow} values.

processes may again be important since the sorbate–sorbent equilibrium is now disturbed, so desorption and even re-adsorption processes might occur.

Presumably river-borne pyrethroids may eventually be exported to estuaries, although few published data exist to test this theory. Indeed, prior to our recent studies, very few published data were available to allow an assessment of even the above sorption processes to be made. It is therefore the purpose of this short review to compile the results of a 5 year programme of research into the sorptive properties of pyrethroids onto minerals, soil and estuarine particles. These have been published previously as individual studies in diverse sources which may not be widely read by sedimentary geochemists. We felt, therefore, that it would be useful to compile such data.

9.2 Materials and methods

The details of the experimental methods have been published in full previously and readers are referred to these studies for a comprehensive description of methodologies (Zhou *et al.*, 1994; 1995a–c). A brief description is given below.

9.2.1 Sorbates

Radiolabelled (^{14}C) synthetic pyrethroids (permethrin, tefluthrin, cypermethrin and lambdacyhalothrin; Zeneca) were purified by radio thin layer chromatography (TLC) or radio high performance liquid chromatography (HPLC) to $> 96\%$ purity and dissolved in hexane as stock solutions.

9.2.2 Sorbents

Three minerals (montmorillonite, aluminium oxide and kaolinite) of known particle size and specific surface areas (Zhou *et al.*, 1994) and a sterilised silty clay loam (Champaign, USA) were used as sorbents, along with suspended particulate matter (SPM) collected in polythene carboys from several sites in the Tees Estuary, UK. SPM was collected by high speed continuous flow centrifugation and stored in 0.1M $NaHCO_3$ at 4°C. Traces of organics were removed from the three minerals by washing with 0.01M NaOH.

9.2.3 Natural organic matter

H^+ saturated humic, fulvic and hydrophilic macromolecular natural organic matter was isolated from the river Dodder, Ireland, by adsorption and cation exchange chromatography (Hayes, personal communication).

Elemental compositions were measured directly on P_2O_5-dried samples (oxygen by difference of ash-free material) and aromaticity by ^{13}C NMR (Zhou *et al.*, 1994).

9.2.4 Adsorption

(a) Natural organic matter onto minerals

Adsorption was carried out in conical flasks or centrifuge tubes. A weighed amount of mineral (20–200 mg) was added to a flask or tube, to which 50 ml of a dilute solution of OM was added. Solutions were adjusted to desired pH, purged with nitrogen, sealed and shaken for 2 days. Solutions were then filtered or centrifuged and supernatant examined by UV–visible spectrophotometry at 350 nm. The amount of OM on the minerals was then calculated by mass balance. Control experiments showed no adsorption to the glassware.

(b) Pyrethroids onto minerals and minerals coated with OM

Sorption was studied by a batch isotherm method using 30 ml centrifuge tubes. Radio-labelled pyrethroid in hexane was added to minerals at the required pH with and without OM (see (a) above). Solvent was removed by nitrogen purge. Tubes were capped, shaken for 24 hr and centrifuged. Radioactivity in the supernatant was measured by scintillation counting following scintillant addition.

Glass-sorbed pyrethroid was measured in the same way after washing off with dichloromethane three times. Control experiments without minerals allowed glass sorption to be independently measured. All experiments were made at least in duplicate. Average recovery was 90%.

(c) Pyrethroid onto soil

Mixtures of radiolabelled tefluthrin (6 μg), water (30 ml) and soil (2.0 g) were placed in centrifuge tubes, shaken (24 hr), the supernatant removed by centrifugation and radioactivity determined as above. The radioactivity of the filtered solids was determined after filtering and treatment with a cocktail specifically formulated to dissolve membrane filters (Zhou *et al.*, 1995c). All experiments were carried out in triplicate and control tubes were used without soil for mass balance and assessment of glass-sorbed tefluthrin.

9.2.5 Desorption

Desorption was studied by three methods but the data from only one are reported here. After adsorption equilibrium, the contents of one tube were poured into a 5 l glass beaker containing 3.5 l of Millipore grade water. The solution was stirred constantly with a magnetic stirring bar. Samples (25 ml) were taken at selected time intervals and filtered through a 0.45 μm membrane filter. The soil particles and filter paper were mixed with 10 ml of a specially formulated scintillation cocktail, shaken and counted. The amount of non-desorbed tefluthrin was calculated.

9.3 Results and discussion

9.3.1 Simple sorption theory

The distribution or partitioning of chemicals between solid and aqueous phases is often described by the *distribution ratio or coefficient* K_D, where

$$K_D = \frac{C_s}{C_w}$$

where

$$C_s = \text{concentration on solid phase}$$
$$C_w = \text{concentration in water phase}$$

However, the C_s value for HOCs is controlled most by the adsorption onto (and/or absorption into) natural organic matter associated with the solid phase.

Thus the organic matter–water (or organic carbon–water) partition coefficient K_{om} (or K_{oc}) is most useful for description of the sorption behaviour of HOCs where:

$$K_{om} = \frac{K_D}{f_{om}} \approx \frac{K_D}{2f_{oc}} \approx 0.5K_{oc}$$

where

$$f_{om} = \text{weight fraction of solid which is natural organic matter}$$
$$f_{oc} = \text{weight fraction of solid which is natural organic carbon}$$

A somewhat related 'surrogate' measure of the partitioning behaviour of HOCs between aqueous and organic phases is the so-called octanol–water partition coefficient, K_{ow}, where:

$$K_{ow} = \frac{C_o}{C_w}$$

where

C_o = concentration of HOC in *n*-octanol phase
C_w = concentration of HOC in aqueous phase

A large database of measured and calculated K_{ow} values is available (e.g. Leo, Hansch & Elkins, 1971) and free energy relationships between K_{ow} and K_{om} have been established.

Readers are referred to the excellent text by Schwarzenbach, Gschwend & Imboden (1993) for a fuller explanation.

9.3.2 Experimental challenge of pyrethroid studies

The extreme hydrophobicity of pyrethroids, as reflected in the high K_{ow} values (Figure 9.1) is reflected in a strong adsorption to many solid surfaces including experimental apparatus such as glass, PTFE and filtration membranes (House & Ou, 1992). Probably this partly accounts for the lack of K_{om} (K_{oc}) values available prior to our studies. Indeed the problem is illustrated by the data shown in Figure 9.2 which shows the adsorption of four pyrethroids, permethrin, tefluthrin, cypermethrin and lambdacyhalothrin onto the glass walls of experimental apparatus at equilibrium. Clearly, in the absence of competing sorbents such as soil, a major proportion ($> 60\%$) of the pyrethroids is not present in free solution but is sorbed to the walls.

Such phenomena dictated two major philosophies for our work with pyrethroids: (a) a mass balance approach was taken to account for all losses in the experiments; (b) as long as the pyrethroids maintained their chemical integrity (did not hydrolyse or otherwise degrade) then use of radiolabelled analytes was desirable. This allowed the low concentrations of the pyrethroids in the aqueous phase to be accurately and reproducibly determined and the mass balance to be computed. The chemical integrity of the radiochemicals was determined by radio-chromatographic techniques (radio-TLC and radio-HPLC).

9.3.3 Adsorption of pyrethroids to minerals and soil

The strong adsorption of pyrethroids for minerals, especially minerals coated with natural organic matter such as are found in soils, is illustrated in our studies by the behaviour of tefluthrin. Thus Figure 9.3 shows sorption of tefluthrin onto organic-free kaolinite, alumina (aluminium oxide) and montmorillonite. Clearly the sorptive capacity of the minerals varies in the order montmorillonite > alumina > kaolinite but all three isotherms are linear over the concentration range investigated. Values of K_D varied from 92 ml g^{-1} (montmorillonite) to 50 (alumina) to 7.6 (kaolinite). This corresponds to only low coverage of the surface of the minerals (Zhou *et al.*, 1995b) and, as expected, also to the specific surface area of the minerals measured by the BET and nitrogen isotherm methods

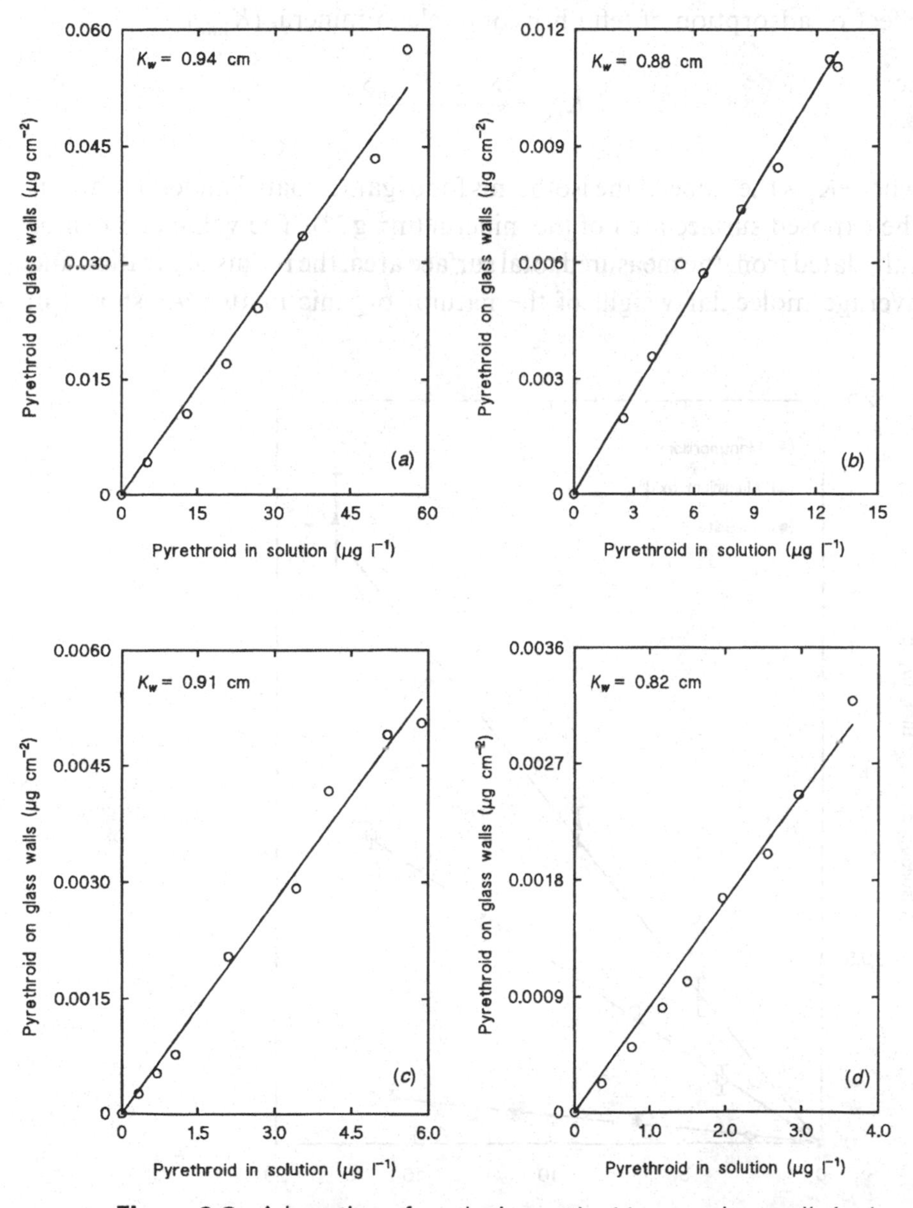

Figure 9.2 Adsorption of synthetic pyrethroids onto glass walls in the absence of sorbents: (*a*) permethrin, (*b*) tefluthrin, (*c*) cypermethrin and (*d*) lambdacyhalothrin.

(Zhou *et al.*, 1994). When the minerals are coated with natural organic matter such as riverine aquatic humic substances, sorption increases several fold due to the strong partitioning of the pyrethroid into the OM (Figure 9.4). The isotherms are still linear ($r^2 > 0.98$) but partitioning increases, with increases in the proportion of organic coatings on the mineral (f_{oc}). The influence of the OM can be obtained by deducting the effect of adsorption of tefluthrin onto clean mineral (K_{min}):

i.e.

$$K_{oc} = \frac{K_t - K_{min}S}{f_{oc}}$$

where K_t is the slope of the isotherms for organic-coated minerals and S is the exposed surface area of the mineral ($m^2\ g^{-1}$). The value of S can be calculated from the measured total surface area, the radius of gyration and average molecular weight of the natural organic matter. As shown in

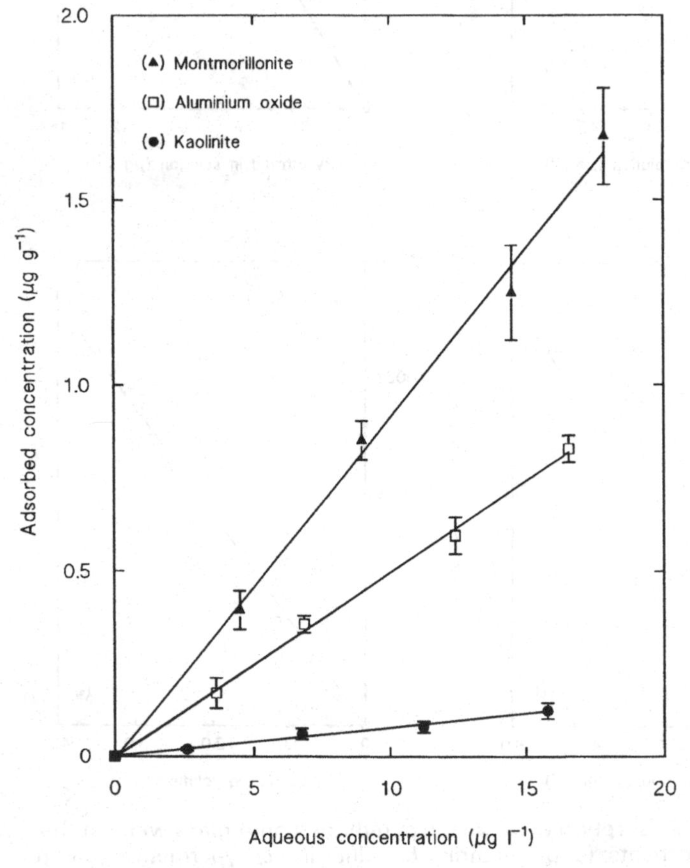

Figure 9.3 Adsorption of tefluthrin by 'clean' mineral particles.

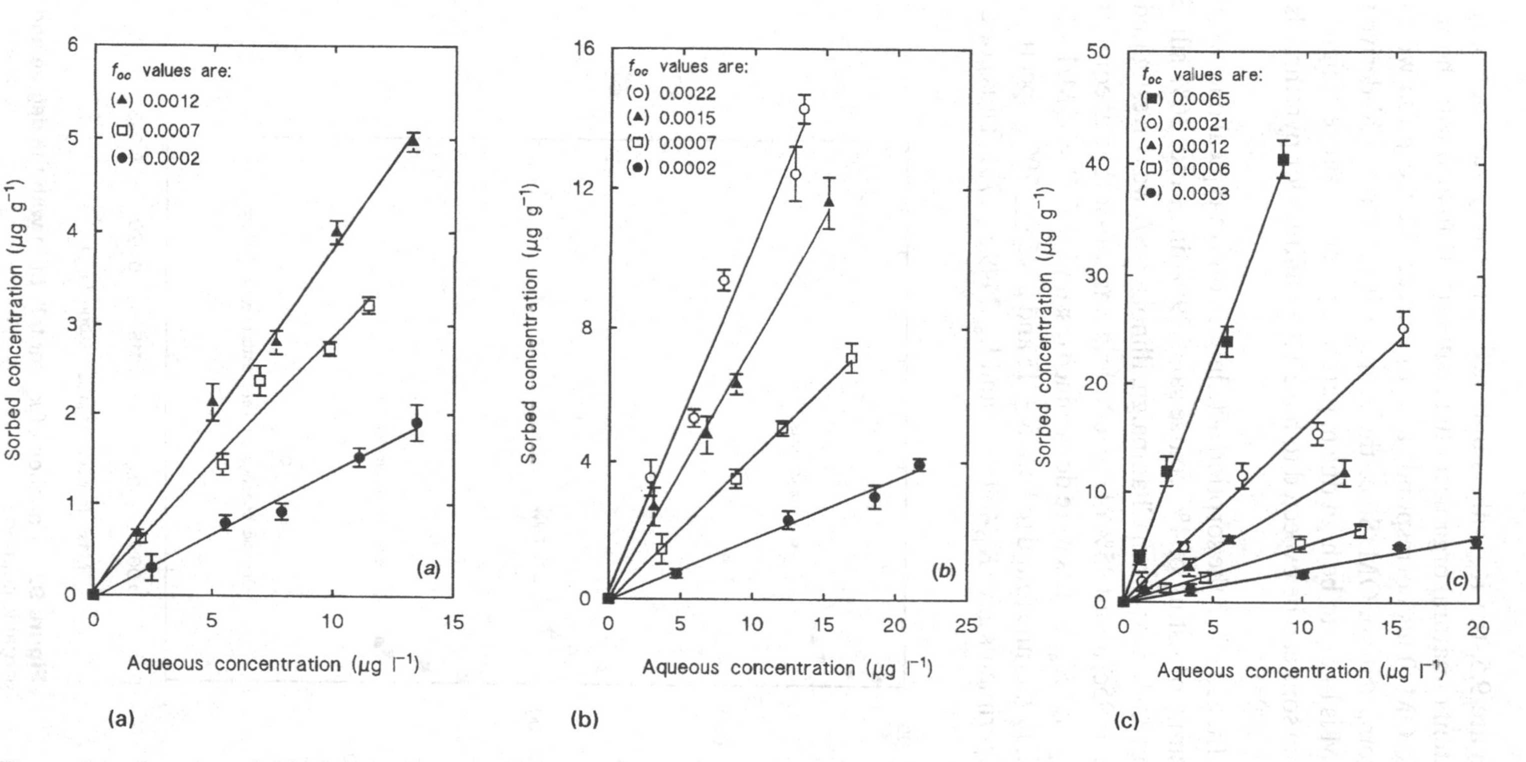

Figure 9.4 Sorption of tefluthrin by montmorillonite coated with *(a)* hydrophilic macromolecular acid, *(b)* fulvic acid and *(c)* humic acid.

Figure 9.5, the K_{oc} values obtained in this way vary not only with the extent of natural organic matter coating (f_{oc}) but also with the character of the OM. This corresponds also to increased adsorption with increased aromaticity of OM. Since the aromaticity of terrestrial-derived (i.e. soil) OM is likely to be high due in part to a preponderance of lignin residues, then soils can be expected to have a high affinity for pyrethroids (Zhou *et al.*, 1994).

Indeed when the sorption of tefluthrin onto a typical mid-west silty clay loam (consisting of 4% coarse sand, 29% fine sand, 36% silt, 31% clay and 3.3% OM) from Champaign, Illinois, USA was investigated (Zhou *et al.*, 1995c) about 95% of the tefluthrin was soil-sorbed at equilibrium. The K_D and K_{oc} values were determined (assuming OC = OM/1.724; Zhou *et al.*, 1995c) and found to be 545 ± 15 and 28454 ± 790 respectively. Repeat experiments gave K_D 469 ± 15 and K_{oc} 24490 ± 761. Thus good precision

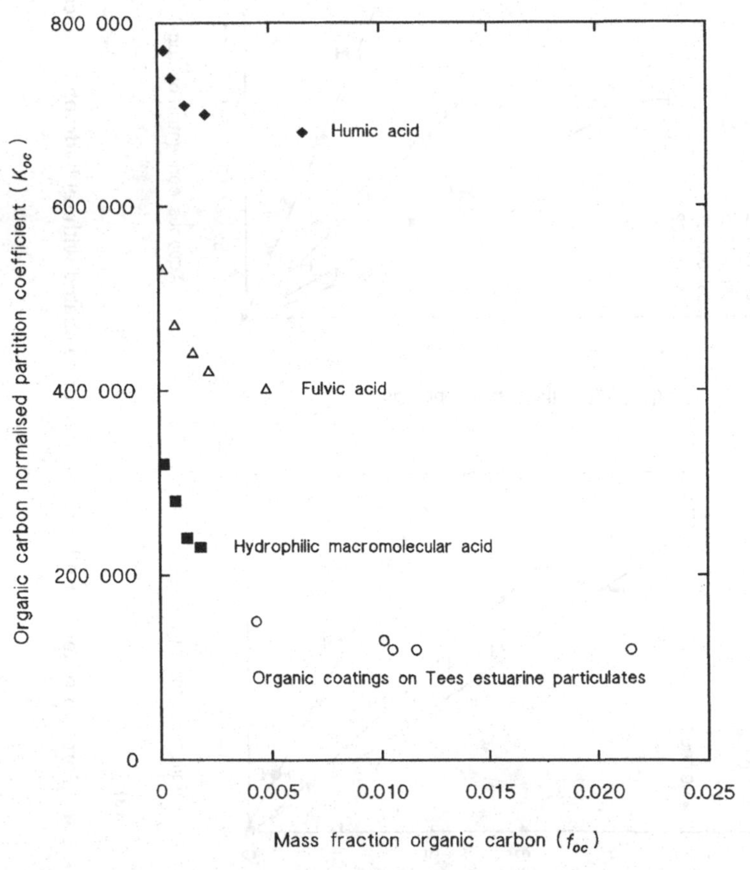

Figure 9.5 Variation of k_{ow} for tefluthrin with the degree and nature of organic coatings.

was obtained (RSD 2.8–3.0%). The lower K_{oc} values obtained for soil (*c.* 27000; $n = 5$) compared with minerals coated with the most aromatic natural OM available to us (humic acids: *c.* 800000) are probably a function of the far higher concentration of soil sorbent used than minerals (the so-called 'sorbent concentration effect'), kinetic effects (Van Hoof & Andren, 1991) and the presence of non-settling colloidal material from the soils which may raise the apparent C_w value and hence lower soil K_D and K_{oc}.

Nonetheless, the soil data show that pyrethroids such as tefluthrin can be expected to be strongly adsorbed to soils. What happens however when this equilibrium is disturbed by physical washing of soil–pyrethroid particles into a river or lake?

9.3.4 Desorption of pyrethroids from soil

Somewhat surprisingly, given its hydrophobicity, our studies suggest that tefluthrin desorbs quite rapidly from soil when presorbed material is transferred to a water body – represented in our experiments by a large (~3.5 l) beaker of fresh water. Indeed, the desorption appears to follow first-order kinetics ($r^2 = 0.98$) with a typical rate constant and half-life of about $0.002\,hr^{-1}$ and 5 hr respectively. Desorption appeared to be independent of pH and salinity of the water but dependent on temperature (all at 95% confidence levels) as shown in Figures 9.6 to 9.8. Calculation of thermodynamic parameters for the desorption process from the van't Hoff equation (Zhou *et al.*, 1995c) showed that the process was exothermic ($\Delta H = -56\,kJ\,mol^{-1}$) and spontaneous ($\Delta G \approx -22$ to $-23\,kJ\,mol^{-1}$). This is consistent with a partition mechanism for tefluthrin (Zhou *et al.*, 1995c).

It is important to note that our experiments were carried out with pure tefluthrin, not the formulated material sold for agricultural use which was not available to us in radiolabelled form. Furthermore, our soil-sorbed tefluthrin samples had not been 'aged'; desorption was studied after only 24 hr adsorption equilibrium.

Nonetheless, these experiments indicate that pyrethroids may re-enter the dissolved phase under some conditions.

9.3.5 Re-adsorption of pyrethroids onto estuarine particles

Assuming that pyrethroids may reach estuaries by riverine transport (budgets of this phenomenon are currently being assessed; LOIS, 1994) the question arises as to whether re-adsorption onto estuarine particulate matter is possible. Given the findings above, this seems entirely reasonable but no published data appeared to exist prior to our studies (Zhou *et al.*, 1995a,b).

Adsorption of tefluthrin, cypermethrin and lambdacyhalothrin onto estuarine particles collected from the Tees is illustrated in Figure 9.9 (Zhou *et al.*, 1995a–c) along with data for the same chemicals sorbed onto hydrophilic natural organic matter coated onto montmorillonite. Values of K_{oc} ranged from 120 000 to about 450 000 for all three compounds depending on the f_{oc} (Figure 9.9). A possible explanation for the decrease in K_{oc} with increase in f_{oc} is that, at the lowest f_{oc} values, all natural OM sites on the particle surfaces are accessible to the sorbate (pyrethroid) but as f_{oc} increases the OM adopts a different interfacial configuration, reducing the availability of sorption sites. Such a phenomenon is not without precedent for other HOCs (Garbarini & Lion, 1986; Schlautman & Morgan, 1993).

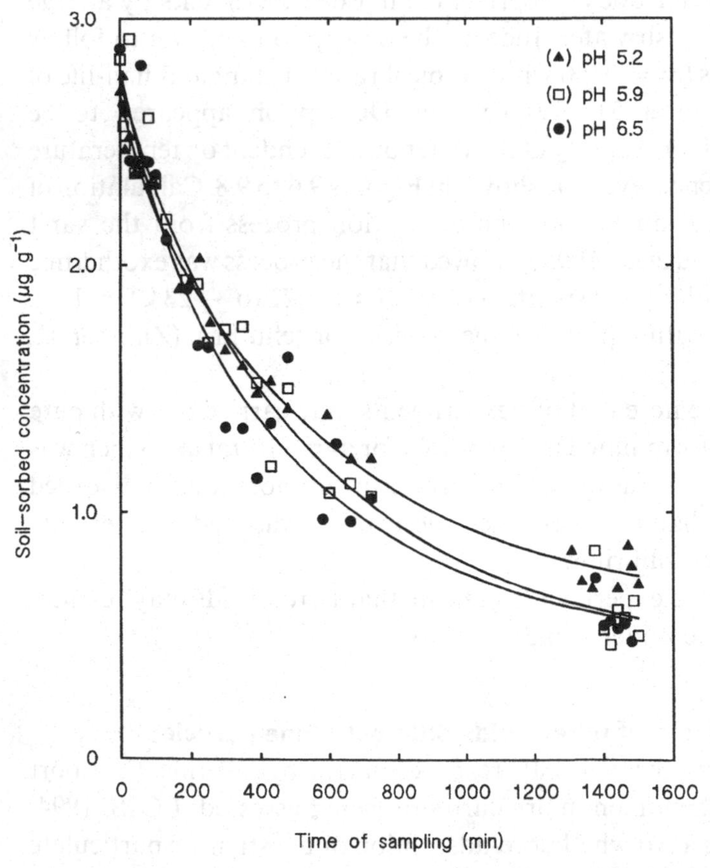

Figure 9.6 Effect of solution pH on tefluthrin desorption kinetics.

9.4 Conclusions

Sorption studies of a variety of widely used hydrophobic pyrethroid insecticides have shown:

(a) that the extreme hydrophobicity of the chemicals requires great care with experimentation in order to quantify losses onto glassware and other apparatus. A mass balance approach to experiments and use of radiolabelled sorbates is recommended.

(b) adsorption of pyrethroids to a variety of minerals is significant, and at the concentrations studied, linear. Values of K_D for clean minerals (K_{min}) ranged from about 100 to about 8 ml g^{-1}.

(c) adsorption to minerals coated with natural organic matter increased severalfold relative to uncoated minerals. Values of K_D for sorption of tefluthrin to a US soil were about 500 ml g^{-1}.

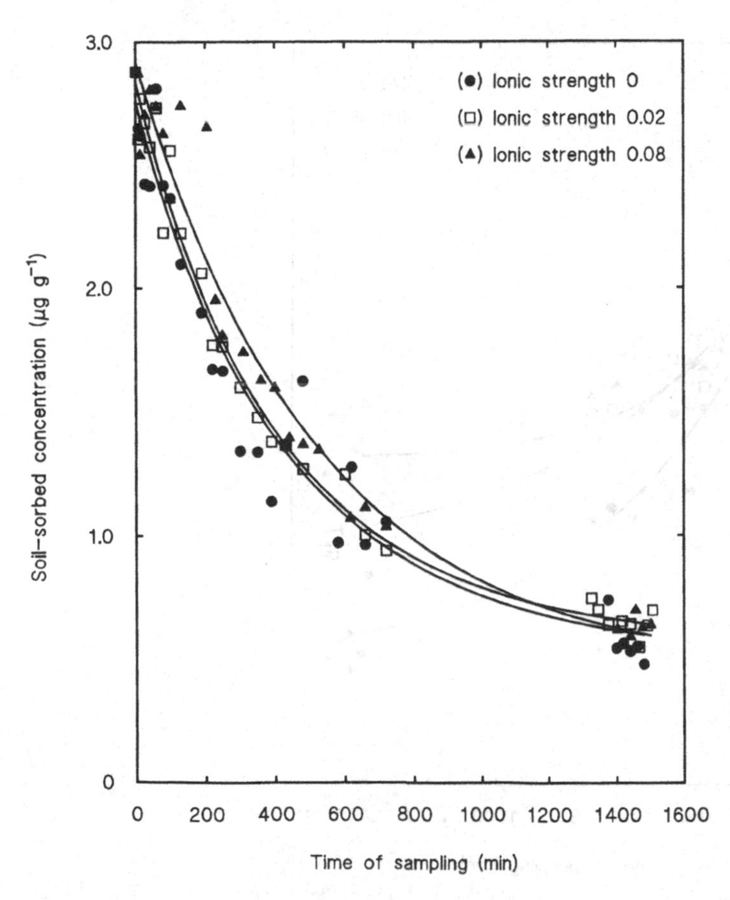

Figure 9.7 Effect of ionic strength on tefluthrin desorption kinetics.

(d) desorption of tefluthrin adsorbed onto soil was rapid and spontaneous ($-\Delta G \approx$ 22–23 kJ mol^{-1}) and followed first-order kinetics ($k \approx 0.002\,h^{-1}$; $t_{\frac{1}{2}} \approx 5\,h$) when the soil was transferred to a vessel containing a relatively large volume of fresh water.

(e) adsorption of a variety of pyrethroids onto estuarine particles from the Tees was strong with K_{oc} values ranging from about 120 000 to 300 000. Values of K_{oc} decreased with increased f_{oc} possibly due to reorientation of the natural organic matter at high f_{oc} values.

Acknowledgements

We are grateful to the publishers of *Water Research*, *Environment International* and the *International Journal of Environmental Analytical Chemistry* and the British Crop Protection Council for permission to

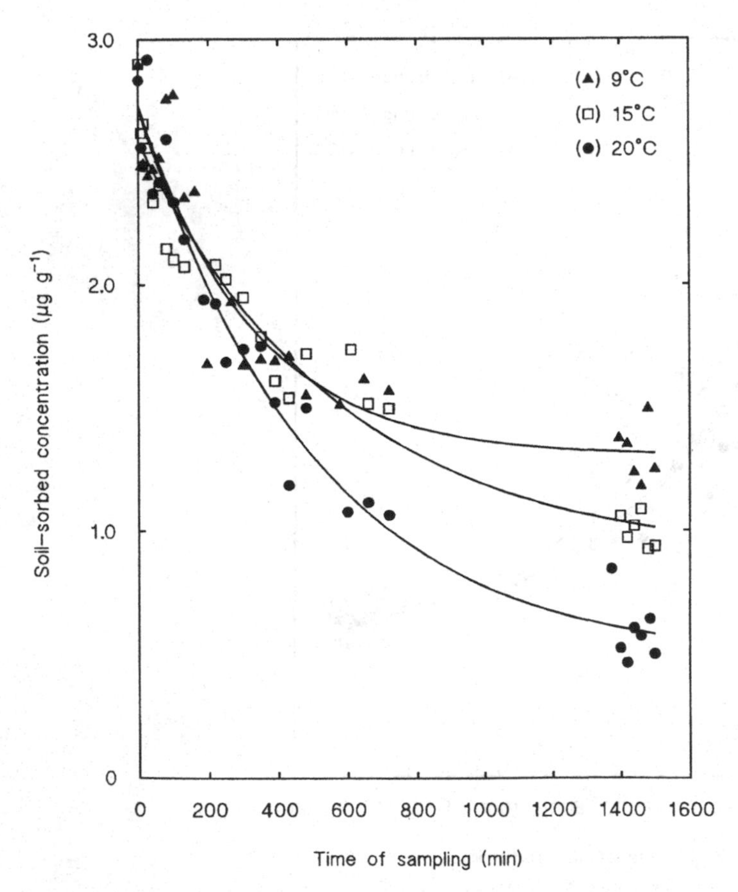

Figure 9.8 Effect of temperature on tefluthrin desorption kinetics.

publish figures from the articles cited. The review presented here represents but a summary of the results of a major input of knowledge and experience from colleagues at Plymouth Marine Laboratory, University of Plymouth, Brixham Environmental Laboratory (Zeneca) and Zeneca Agrochemicals Ltd, UK, especially Professor R.F. Mantoura, Dr J. Braven, Brian Harland and Mike Lane. We are grateful to Zeneca for funding the research and for permission to publish. We are also grateful to Dr Mike Hayes, University of Birmingham, for the supply of natural aquatic humic material isolated from the River Dodder, Ireland.

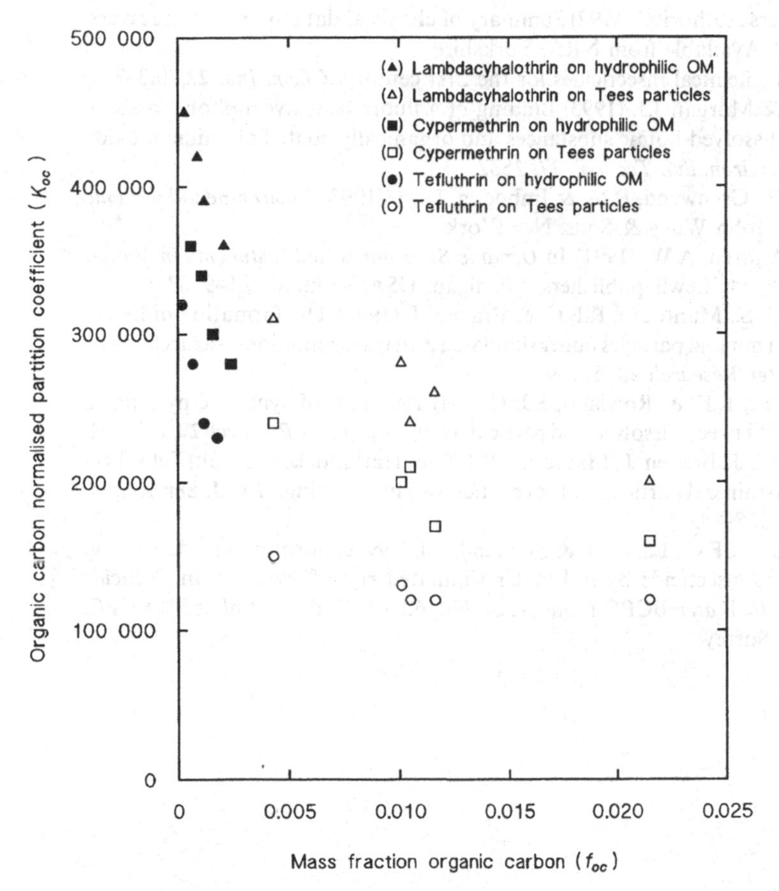

Figure 9.9 Summary of the organic carbon normalised partition coefficients for pyrethroids.

References

Garbarini, D.R. & Lion, L.W. (1986) Influence of the nature of soil organics on the sorption of toluene and trichloroethylene. *Environ. Sci. Technol.* **20**, 1263–9.

Hill, I.R. (1985) Effects on non-target organisms in terrestrial and aquatic environments. In *The Pyrethroid Insecticides*. Leahey, J.P. ed. Taylor & Francis, London. 151–262.

Hill, I.R. (1989) Aquatic organisms and pesticides. *Pest. Sci.* **27**, 429–65.

House, W.A. & Ou, Z. (1992) Determination of pesticides on suspended solids and sediments: investigations on the handling and separation. *Chemosphere* **24**, 819–32.

Leo, A., Hansch, C. & Elkins, D. (1971) Partition coefficients and their uses. *Chem. Rev.* **71**, 525–53.

LOIS (1994) The first LOIS RACS (C) workshop. Burwalls, Bristol, 28–30 March 1994. Copies from P. Donkin, J. Harris, P. Watson or R. Willows. Plymouth Marine Laboratory, Plymouth, Devon.

NRA (National Rivers Authority) (1992) Summary of chemical data for Yorkshire rivers, 1990–1992. Available from NRA, Yorkshire.

Perrior, T.R. (1993) Chemical insecticides for the 21st century. *Chem. Ind.* **22**, 883–7.

Schlautman, M.A. & Morgan J.J. (1993) Binding of a fluorescent hydrophobic organic probe by dissolved humic substances and organically coated aluminium oxide surfaces. *Environ. Sci. Technol.* **27**, 2532.

Schwarzenbach, R.P., Gschwend, P.M. & Imboden, D.M. (1993) *Environmental Organic Chemistry*. John Wiley & Sons, New York.

Van Hoof P.L. & Andren, A.W. (1991) In *Organic Substances and Sediments in Water*. Baker, R.A. ed., Lewis publishers, Michigan, USA. Volume 2, 149–67.

Zhou, J.L., Rowland, S., Mantoura, R.F.C. & Braven, J. (1994) The formation of humic coatings on mineral particles under simulated estuarine conditions – a mechanistic study. *Water Research* **28**, 571–9.

Zhou, J.L., Mantoura, R.F. & Rowland, S.J. (1995a) Partition of synthetic pyrethroid insecticides between dissolved and particulate phases. *Water Research* **29**, 1023–31.

Zhou, J.L., Rowland, S.J., Braven, J., Mantoura, R.F.C. & Harland, B.J. (1995b) Tefluthrin sorption to mineral particles: role of particle organic coatings. *Int. J. Env. Analyt. Chem.* **58**, 275–85.

Zhou, J.L., Mantoura, R.F.C., Lane, M. & Rowland, S.J. (1995c) Sorption and desorption of tefluthrin insecticide by soil under simulated run-off systems. In *Pesticide Movement to Water* BCPC monograph No. 62. (A. Walker *et al.*, eds.) BCPC, Farnham, Surrey.

Index